Boll Weevil Blues

Boll Weevil Blues

COTTON, MYTH, AND POWER
IN THE AMERICAN SOUTH

James C. Giesen

The University of Chicago Press   CHICAGO & LONDON

James C. Giesen is assistant professor of history at
Mississippi State University.

The University of Chicago Press, Chicago 60637
The University of Chicago Press, Ltd., London
© 2011 by The University of Chicago
All rights reserved. Published 2011
Printed in the United States of America

20 19 18 17 16 15 14 13 12 11   1 2 3 4 5

ISBN-13: 978-0-226-29287-8 (cloth)
ISBN-10: 0-226-29287-8 (cloth)

Library of Congress Cataloging-in-Publication Data
Giesen, James C. author.
Boll weevil blues : cotton, myth, and power in the American South /
James C. Giesen.
pages cm.
Includes bibliographical references and index.
ISBN-13: 978-0-226-29287-8 (alk. paper)
ISBN-10: 0-226-29287-8 (alk. paper)
1. Boll weevil—Economic aspects—Southern States.   2. Cotton—Economic
aspects—Southern States—History—20th century.   3. Southern States—
Economic conditions—20th century.   I. Title.
SB945.C8G54 2011
338.1'73510975—dc22
2010047834

CONTENTS

In December 1904, President Theodore Roosevelt addressed Congress on the state of the union. Among the most important threats to the nation's health and security, Roosevelt cited "a Central American insect that has become acclimated in Texas and has done great damage." U.S. Department of Agriculture (USDA) scientists had identified the cotton boll weevil, a small beetle that feeds on the squares of the cotton plant, in Texas less than a decade earlier, but as evidenced by the special attention the president was giving the insect, it had already become a national issue. Roosevelt cautioned that "the boll weevil is a serious menace to the cotton crop." The same year, agents for the state of Georgia had become so concerned about the weevil, which was still hundreds of miles from their state's border, they banned the importation of any cottonseed or lint cotton from weevil-infested Texas or Louisiana. Officials were concerned the bugs might stow away in a container of seed and prematurely invade their state's cotton fields.[1]

Their caution was well founded. By 1904, people across the globe knew of the boll weevil and its threat to the world's cotton supply. Articles about the insect had appeared in national periodicals as well as in innumerable local newspapers; they were stark in tone.[2] "We know of no single subject that contains more of importance *to the entire country's economic interests* than the devising of measures to arrest and, if possible, eradicate this scourge," argued a 1904 article in *Science*.[3] Later that year, David F. Houston, president of Texas A & M College, wrote that the boll weevil "constitute[d] the greatest menace to cotton production that the cotton farmer has had to face," and that attempts to stop the pest were "useless." USDA special agent Seaman Knapp

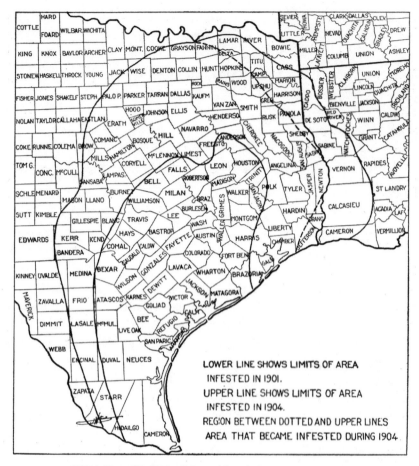

FIGURE 1. USDA Map of Boll Weevil Spread (1904). From W. D. Hunter, "The Status of the Mexican Cotton Boll-Weevil in the United States in 1903," in USDA, *Yearbook of Agriculture*, 1903 (Washington, D.C.: Government Printing Office, 1904), 206.

had visited weevil-plagued Texas the previous year and later described "a wretched people facing starvation" and "whole towns deserted."[4]

Accompanying many of these print reports were maps of the pest's advance, showing menacing-looking waving bands of destroyed territory. Readers could easily envision the pest soon engulfing the entire cotton South. The USDA and state agriculture departments originally prepared these surveys for pamphlets sent to farmers and researchers; but when the maps were stripped of the dry, informative context that surrounded their original publications and were reprinted in papers across the country, millions of readers gained the impression that the pest had completely covered areas of the South and that cotton growing had ceased within them.[5] (See Figures 1–3.)

In addition to the frightening tales of the weevil that appeared in print, by 1904 southerners were hearing (and telling) alternative stories in song and conversation. Traveling singers in Texas had already written a number of boll weevil songs, and the stories they told of the pest's invincibility spread the South over and did so even more quickly than the weevil itself. Before long, the boll weevil had become much more than an entomological affliction; the insect transformed into nothing less than a symbol of the South's long history of rural poverty, racism, and environmental blight.

Between the 1890s and the early 1920s the boll weevil slowly and unevenly crawled and flew across the cotton South from Texas to the Atlantic Ocean, and by any measure it destroyed a lot of cotton along the way. In the 1890s, some Texas counties reported 70 percent losses. The weevil took 19 percent of the Lone Star State's entire 1903 cotton crop, even though the pest was present only in about half of Texas counties. Louisiana, Arkansas, and Mississippi experienced similarly striking crop losses as the weevil moved through to the north and east. No state suffered a single-season setback like Georgia; it lost 45 percent of its 1921 crop to the bug, but individual counties

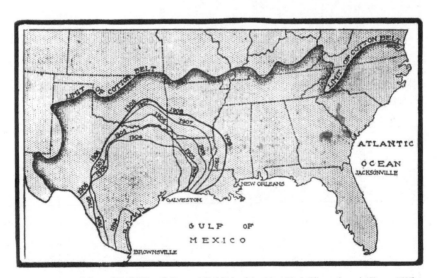

FIGURE 2. Map of Boll Weevil Spread Published in *New York Times* (1910). From "Why the Deadly Boll Weevil, Bringing Revolution With Him, Is Called the 'Prosperity Bug,'" *New York Times*, January 9, 1910, part 5, p. 13.

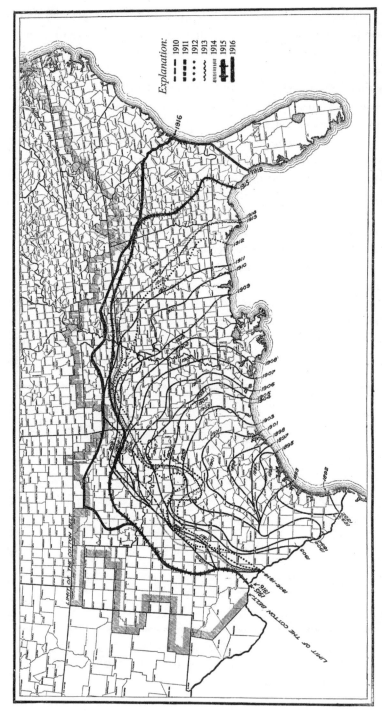

FIGURE 3. USDA Map of Boll Weevil Spread (1922). Walter D. Hunter and B. R. Coad, "The Boll Weevil Problem," *USDA Farmers Bulletin* 848 (1917).

experienced much higher damage. By the time the pest reached the limit of the cotton belt, it had destroyed an unseemly amount of the region's cotton, a crop that served both as the South's principal economic engine and as a crucial cultural linchpin. At the end of the twentieth century, researchers estimated that the beetle had destroyed *tens of billions* of pounds of cotton since its arrival in the United States, the value of which approached one trillion dollars. These mind-boggling numbers suggest not only cotton loss, but also radical changes in the fabric of southern economic and social life.[6]

The historical record seemingly agrees. In the early twentieth century it was easy to find people of all stripes raving about the bug. Harris Dickson claimed the weevil "marched through Georgia like Sherman to the sea, and creating far more havoc."[7] A writer for *Harper's* concluded in 1925 that the pest "cost America her cotton dominance of the world."[8] An entomologist dispatched by the USDA to research the weevil wrote that in parts of the South, "planters and speculators committed suicide" rather than face the beetle's wrath.[9] These are common appraisals of the pest's effect on the South from the 1890s to 1930. Newspapers, entomologists, family storytellers, traveling country and blues singers, tenant farmers, and planters have presented the boll weevil as a wrecker of plantation agriculture, an unstoppable natural disaster that swept through the South unchecked and brought ruin to its people.

This book argues that these staggering figures of crop losses, and the social and economic upset they triggered, were only part of the insect's impact on the cotton South. It was the *idea* of the boll weevil, more than the physical destruction it wrought, that most profoundly changed the region. Contrary to the narratives offered in songs and stories from the early twentieth century, and even the powerful argument made by those staggering statistics of cotton loss, the boll weevil did not forever change the rural South. The pest's greatest consequence was not the many stands of cotton it devoured, but rather the great explanatory power that people found in the weevil. From the moment laborers discovered the insect in a field near Corpus Christi, there was dissonance between the physical damage the boll weevil inflicted and the significance people attached to it.

The weevil's destruction of southern cotton is a big part of the story this book tells, but it also focuses on the collective anxiety produced by the *perception* that the weevil's crop destruction would bring social and economic ruin to the South, and the transfer of that notion from a remote cotton field near the Mexican border to the halls of the agriculture department in Washington, to state capitols, to juke joints, to mansion porches, and to country stores. It will argue that the history of the weevil's spread across the region is as much about fear and expectations as it is about reduced crop yields and their fiscal consequences. The movement of this fear, as much as the movement of the weevil itself, is the topic of this book.

From the 1890s to the Great Depression, the boll weevil myth—told and retold in various forms by people from all parts of society—was this: an enormous, amorphous swarm of insects was marching inexorably across the South posing the greatest threat to southern life since the Civil War. Without a radical rethinking of the basic ways of life of the region, the myth argued, it would bring—indeed, by the time President Roosevelt gave his State of the Union message it was already bringing—economic and social revolution. It was the weevil's threat that forced farmers, merchants, mill owners, politicians, and countless others to rethink their relationship with the cornerstone crop of the region's economy and with the soil in which it was planted, the laborers who worked it, and the natural environment that fostered it. More than a mandate for change, this cotton bug provided for a range of southerners an easily digestible, quick, powerful excuse for almost all of the problems facing the rural South in the first half of the twentieth century. According to the myth, it was not landowning patterns, antiquated credit systems, white supremacy, or declining soil fertility that made the rural South what it was, it was an "Act of God," a voracious natural enemy. A range of people across race and class lines attempted to use fear of the weevil as both a powerful excuse for the South's poor conditions and as an answer to its most vexing problems. In short, a boll weevil myth was born, one that contained an explanation for and a solution to a variety of southern troubles.[10]

This notion that that the boll weevil had the power to fundamentally alter the course of the region came from a broad swath of southern society and made its way into all corners of southern culture. Large plantation owners, tenant farmers, rural researchers, local and national politicians, bankers, and merchants were all scared by the weevil, but each thought that its damage to cotton would mean something different. As the chapters that follow explore, southerners, along with many people from outside of the region, projected onto the impending devastation not only the consequences they feared, but also outcomes for which they yearned. Men and women passed on their weevil myths in political speeches, private letters, songs, family tales, and conversations on street corners in order to bend the future of the rural South toward their own visions of economic and social destruction and salvation.

While myths, rumors, and misinformation often arise out of crises, especially environmental ones, this divergence between the perception and reality of the boll weevil's impact on the South was, in most cases, not an accidental creation born of a moment of emergency.[11] After all, the weevil's transverse of the American South was no quick thing; there wasn't the urgency of a river cresting over a levee, a hurricane spinning toward landfall, or a fire moving wildly toward a community. The insect took decades to get from Texas to Virginia. Most people had time to try to make sense of the impending disaster and to capitalize on the change that it promised to bring.

Southerners and nonsoutherners alike used this threat to push their own various reform agendas. Some sharecroppers were encouraged by the pest's disruption of the cotton economy and hoped to profit from the fear it produced by gaining access to landownership. Agricultural scientists saw in the insect an opportunity to advance their own research and garner funding and resources for their institutions. For example, years after trumpeting the weevil's threat to southern cotton, Theodore Roosevelt hailed the weevil as a "blessing in disguise" because of the modernization and crop diversification it was forcing on farmers. Like Roosevelt, people both inside and outside of the cotton regions understood not only the insect's threat, but also the great possibilities that the weevil offered the rural South. To hear these and other observers tell it, not only had the pest leveled fields, it had upset seemingly unassailable institutions of southern life. No matter the angle of their conclusions, these accounts presented the pest as a transformative force on life in the southern countryside.[12]

The chapters that follow trace the trek of the boll weevil and its legend across the South from the 1890s to the late 1920s and keep as their focus counties, fields, and towns, rather than the region as a whole. It is on this local level that the complex story of the boll weevil invasion and the creation of the attendant myths, as well as the full range of effects created by the boll weevil, reveals itself. The experiences of four specific subregions — east Texas, the Yazoo-Mississippi Delta, southeastern Alabama, and central Georgia — form the core of this analysis. In each of these areas cotton meant something different to the economy and society, in part because of substantial differences in climate, soil, topography, weather, and other environmental factors. Each place had its own social, economic, and political history related to those environmental forces as well. Further complicating the story was the timing. The boll weevil arrived at different historical moments for each place; in the western South, cotton was a new crop, while in the eastern states it had been in a long decline. Thus, the physical environment, labor relationships, landownership, and political and class divisions together tempered the boll weevil's impact in each place, and the powerful, transformative images of the pest developed differently as a result.

*

This intersection of the boll weevil, local power, and myth is one never before explored by historians. Scholars of the rural South have discussed the boll weevil in a variety of often-inconsistent ways. The pest makes an appearance in nearly every general history of the New South period, and in more narrow examinations of economics, migration, farming, black rural life, technological change, and the growth of the state. While it is not hard to find a mention of the boll weevil in works of American history, it is hard to find close analysis of it. Despite its relevance to southern historiography, there has been no monograph devoted to the pest.[13] Those who have studied the pest, however, have resisted the temptation to endorse the myth of the weevil's transformative power. In fact, most of the major recent scholarship on the New South and southern agriculture presents the boll weevil as an important factor in the region's history, but not a singularly powerful one.[14]

This book argues that explaining that the boll weevil did not single-handedly destroy or re-create the rural South does not tell the whole story. Historians have too often understood the weevil's history as simply one of man versus nature, a clash of chemicals and state policies. Scholars have failed to explain the rise and power of the pest's myth, which leaves out a crucial aspect of why it has been so important to southern life. None of the previous scholarship explains how these ideas about the weevil either reflected or challenged local conceptions of nature, agriculture, race, or class. One factor in the oversight of culture in previous work has been its narrow focus on farming. Traditional southern agricultural history has focused on the rise of land grant colleges, the USDA, the extension service, and state departments of agriculture, and has relied too heavily on the records of these organizations as primary sources. In regard to the boll weevil, this has sometimes meant that historians end up parroting the party line given by extension agents, scientists, or planters, groups that benefited from presenting the boll weevil as a greater threat than it was. Historians should understand the scientists who studied the weevil as part of a "political entomology," a term historian Karen Brown coined to describe how South African insect researchers had their work mitigated by the social and economic structures of society.[15] While I too have interrogated the records left by these farm organizations, I have done so with an eye toward myth creation and have also worked to balance this perspective with evidence offered by sharecroppers, small farmers, and others whose voices are less easily heard.

For this reason, even as I spend the pages that follow charting the pest's destruction of cotton, cultural interpretations of the boll weevil lie at the heart of my argument. While many historians have used stanzas from a boll weevil song anecdotally, or mentioned the boll weevil statue in Enterprise, Alabama, in passing, few have seriously considered the meaning of these songs—or the monuments, poems, and stories devoted to the weevil—and no work has taken seriously the origin and transformation of these cultural expressions.[16] This book attempts to better connect cultural representations of the insect to the reality of its arrival in southern cotton fields, and to explain how these ultimately false portrayals of the boll weevil helped to create an image of the rural South as a depressed, deserted, and benighted landscape.

Another way that I hope to change the understanding of the boll weevil's effects on the South is by approaching the insect not just as an agricultural pest, but also as an environmental disaster. I argue that considering the boll weevil as both an entomological and a mythical threat forces us to rethink what we mean when we call something a natural disaster. Scholars pointed out long ago that disasters society labels "natural" are very often human-induced, or at least aided by humans' interaction with the environment. To use the most famous example from environmental history, farmers in the Great Plains helped create the Dust Bowl by trying to force the land to produce crops on an industrial scale.[17] Economics compelled farmers to change the land; drought and social disaster followed. The case of the boll weevil, however, is one in which the destruction of vast quantities of cotton did not immediately or permanently alter the agro-environment of the South. The boll weevil teaches us that not only can men and women create these disasters, they can effectively resist them as well, or at least can protect social structures like class and race in the face of these "natural" threats. In fact, nature itself, in the form of the boll weevil, became a weapon with which to resist the very change that its appearance threatened to cause. Planters and sharecroppers, country store managers and mill owners, local extension agents and directors of the USDA all used this environmental threat to either test or reinforce traditional class and racial orders.[18]

Finally, beyond the boll weevil story, I am seeking to delve into the intersection of ideologies about the environment with race, class, and geography in a way that expands our understanding of both agricultural and environmental history. This work attempts to broaden historians' view of cultural interpretations of the environment by explaining not only how and why humans developed new understandings of the natural world, but also the ways these constructions shaped much broader understandings of southern history. The history of this beetle's long movement, both literally and figuratively, from entomological nuisance to national natural disaster is at its essence a story about the intersection of environmental, economic, and social power and the ideologies that people develop to either sustain or thwart it.

CHAPTER 1

Myth Making on the Cotton Frontier

In early 1903, farmers in Terrell, Texas, were nervous about the slowly spreading boll weevil. The town's business leaders called a meeting of the area's most influential farmers and invited famed agricultural educator and USDA special field agent Seaman Knapp to address them. The old professor later claimed that he was there to allay farmers' fear of the encroaching pest, to put at ease a community demoralized by the prospect of the boll weevil's arrival. But his practical advice for farmers could not counter their hysteria. According to Knapp, when he concluded his remarks, a man sitting in the back of the room rose to his feet and "stated that it was impossible to fight the weevil." The pest "was proof against everything that had been tried," the farmer claimed. The man explained that only a few days earlier, he had captured a few live weevils and put them in a jar of "ninety-five per cent pure alcohol." Four hours later he poured out the jar and the bugs emerged alive, "only staggering drunk." Dumbfounded, the farmer collected the inebriated insects, "sealed them in a tin can, threw them into a brush heap and set it on fire." A few minutes later, he watched as "the solder melted and the red-hot weevils flew out and set the barn on fire." Knapp knew that in this apocryphal story lay an important truth: the boll weevil had caught the attention of Texas farmers like nothing else ever had. In this bug, and more precisely in people's fear of it, Knapp recognized, was contained the power to transform southern agriculture forever.[1]

By 1903, the boll weevil had been present in Texas for at least ten years, but mythical stories of the pest's formidable strength and assured destruction of cotton life were everywhere. As Knapp traveled east Texas that year, he talked to farmers, bankers, and merchants about the boll weevil's advance and surveyed the pest's impact on rural life firsthand. Later he recalled, "I saw hundreds of farms lying out; I saw a wretched people facing starvation; I saw whole towns deserted; I saw hundreds of farmers walk up and draw government rations, which were given to them to keep them from want." According to Knapp, the boll weevil had destroyed the cotton way of life in Texas.[2]

But had it? There can be no doubt that the boll weevil hit Texas hard in the late nineteenth and early twentieth centuries. It destroyed thousands of tons of cotton. The pest disrupted the lives of tenant farmers and landowners, hurt railroads, merchants, and banks, and became the obsession of state politicians and scientists. However, interpretations like Knapp's overplay the extent to which the pest affected the economic and social systems girding cotton-farming communities. In fact, Knapp and his colleagues came to realize soon after the pest's identification in Texas that the boll weevil's invasion presented an unmatched opportunity to rethink the agricultural systems of not only the Lone Star State, but of the entire South.

In spite of this evidence of the boll weevil's destructiveness, cotton production actually increased in Texas during this period. Perhaps even more surprising, cotton growers barely altered their farming methods when the insect appeared. This seeming paradox of the destruction of Texas cotton and the persistence and extension of cotton production in the state suggests not simply a popular misunderstanding of the boll weevil's history in the state; it points to the nascent development of a lasting myth about the pest and its effect on rural Texas.

The boll weevil myth born in Texas grew from the bona fide threat the pest posed to cotton culture. Landowners, tenant farmers, state and federal politicians, bureaucrats, and scientists understood the boll weevil threat differently, and offered their own interpretations of the bug's meaning for the South, but it was federal officials like Knapp who had the most success shaping the myth and using it to build private and public support for their work. These state and federal employees saw in the pest a solution to the South's overdependence on cotton and a way to overcome farmers' unwillingness to employ modern, scientific farming methods. Most importantly, these farm educators and scientists knew the pest could foster relevance, even need for their institutions: the USDA, state agricultural departments, land grant colleges, and research farms. To be sure, these experts wanted to limit the destructiveness of the boll weevil, but they also sought a more central place in American farm life.

*

Long before the arrival of federal investigators, the boll weevil roamed the Texas borderlands unknown, or at least unrecognized. Before the cotton pest became the cause célèbre of southern reformers, it was a bug in a field that no one could identify. Farmers probably talked to one another about the pest; landowners surely asked their tenants about the little brown beetles. But on the diverse, ever-changing, chaotic borderlands of southern Texas in the late nineteenth century, there was already enough for farmers to worry about, as the pest was just another factor in the gamble that was cotton farming.

The Texas borderlands had long been home to speculators, drifters, and those in search of the safety provided by remoteness. Fifty years earlier, American slaves and Mexican *peones* sought the border as a place of refuge, but by the 1880s, the region was undergoing a radical restructuring at the hands of the cotton plant and its social, economic, and environmental impacts. The fleecy white crop had found a home in east Texas decades earlier, but as available land there became scarce and extant cotton land grew less productive, farmers pushed the crop to the west and south toward the Rio Grande. White, black, and Mexican workers from both sides of the border followed the plant into southern Texas to work these newly transformed fields. The arrival of cotton to the region not only brought changes in labor, economic, and social systems, it connected this plant—the most important cash crop grown in the United States—to the habitat of its fiercest predator. When Texans extended the Cotton Belt to the Rio Grande, they in essence made a connection between the boll weevil habitats of northern Mexico and coastal Texas into a two-thousand-mile cotton buffet stretching to the Atlantic Ocean.[3]

Though traditional accounts of the boll weevil's arrival in the United States begin with the USDA's formal identification of the insect in 1894, recent scientific literature suggests that the pest was present in Texas for decades earlier. The boll weevil's principal plant host is commercial cotton (*Gossypium hirsutum*), but it can sustain itself, and in some cases thrive, on a small number of alternative plants. One of these potential hosts is a close relative of cotton, *Cienfuegosia drummondii*, which grows natively in only one area of the United States, the coastal region of south Texas near the Mexican border. The very place where the boll weevil first appeared in commercial cotton is, not coincidentally, home to the insect's second most important host plant. In the nineteenth century, botanists and entomologists cared little about this flowering plant or the bugs that took their nourishment from it. In fact, no one seems to have identified it as a host for the boll weevil until the mid-twentieth century.[4]

But when does a pest become a pest? In the 1890s, *Cienfuegosia drummondii* escaped professional examination because it grew in an ecology removed from what most Texans or Mexicans understood as economically or socially important, yet the plant was within range of the migration routes of weevils living in Mexican cotton fields. As a result, the boll weevil was almost certainly present in south Texas long before area farmers noticed it destroying their commercial cotton crop. Even later, after Texans remade the grazing land that was nearer this alternative weevil habitat into cotton fields, the weevil certainly bothered farmers before they knew what it was. That the boll weevil was present in the American South earlier than southerners, as well as historians, have understood, reveals the privileged role that the government, and the USDA in particular, played in shaping the traditional narrative of the boll weevil's effect on the Cotton Belt. That scholars have begun their stories with the government's arrival in south Texas is evidence in itself of the success of the USDA and its related agencies to write the narrative of the boll weevil and to create its own image as a bulwark against this "revolutionary" natural enemy.[5]

It was not until 1894 that the insect came to the attention of an American outside of south Texas. That year Charles H. DeRyee, a druggist from Corpus Christi, sent a few dead specimens to the Commissioner of Agriculture in Washington, D.C. His accompanying letter advised that the "cotton in this section has been very much damaged and in some cases almost entirely destroyed by a peculiar weevil or bug which by some means destroys the squares and small bolls." The message touched off a surprising amount of activity within the USDA. Leland O. Howard, official entomologist for the department, examined but could not identify the samples. He passed the specimens on to other experts both inside and outside the department. When several renowned American entomologists failed to identify the bug, the department finally sent a sample to Paris, France, where the entomologist August Salle identified the insect as *Anthonomus grandis*, the cotton boll weevil.[6]

Salle based his identification on the classification Swedish entomologist C.H. Boheman made fifty years earlier. In the mid-nineteenth century, Boheman had examined boll weevils collected near Vera Cruz, Mexico. He recorded few details about the weevil aside from its appearance. No one seems to have tracked the insect's movement during the mid-nineteenth century until German entomologist Eduard Suffrian noted its presence in Cuba in the early 1870s. A decade later, British-born entomologist Edward Palmer, working for the USDA, traveled throughout Mexico and Central America observing insect life. He recorded the ravages of a "small, dark-colored weevil" near Monclova, Coahuila, Mexico, roughly 115 miles from the U.S. border. Palmer had written to the U.S. Commissioner of Agriculture to report that infestations of this pest had grown so extreme in parts of northern Mexico that farmers had abandoned cotton altogether. The USDA made no effort to follow up on Palmer's research, however. The Mexican and American cotton districts were still unconnected at that point, and the entomological community knew nothing of the pest's range of migration or the host plants available in south Texas. In the final two decades of the nineteenth century, Texans planted cotton closer and closer to the border, however, bringing the American cotton belt ever closer to Mexico and the boll weevil.[7]

After the USDA identified the weevil specimens sent from Texas, the organization began a remarkable period of activity. First, the department sent word back to Charles DeRyee, warning the Corpus Christi businessman of the "imminent danger that [the weevil] may spread into other portions of the cotton belt." Second, with a quickness that indicates the USDA's relatively small size and bureaucratic agility at the time, entomology chief Howard dispatched Tyler Townsend to south Texas to investigate the weevil damage. Townsend, a department entomologist stationed at Las Cruces, New Mexico, had recently returned from a trip to northern Mexico to research cotton pests.[8]

In November 1894, Townsend arrived in Eagle Pass, Texas, a border town southwest of San Antonio, and began talking to local farmers and examining weevil specimens. From Eagle Pass, he traveled east to Corpus Christi and Brownsville, where locals told him that the weevil had been present for at least a decade, and into northern Mexico, in hopes of finding a way that local farmers successfully combated the bug. Townsend remained in the region for a month, talking with landowners, measuring the decline in the cotton crop, and observing the life cycle and habits of the beetle. In December, he returned to New Mexico and prepared his report.[9]

Townsend based his initial recommendations on a close study of the physiology of both cotton and the boll weevil. It was clear to the entomologist that the combination of the weevil's dependence on cotton and the manner in which it destroyed the plant made it a profound threat to future cotton cultivation anywhere the crop grew. Most important of the discoveries made by Townend, as well as scientists who have since investigated the pest, was the realization that the boll weevil's entire life cycle is centered around the cotton plant. Individual weevils feed on the plant's fibers, lay eggs in its squares,

FIGURE 4. Adult weevil, magnified 140 times. Photograph by Carter D. Poland of Anniston, Alabama. Carter D. Poland Photograph Collection, Record Group 268, Auburn University Special Collections and Archives.

grow in its enclosed buds, and hibernate on the edges of its fields. Of the weevil's four life stages (egg, larva, pupa, and adult), scientists have come to understand, it lives three inside the cotton plant itself. As Townsend discovered, this dependence on the crop is the principal reason for the weevil's destructiveness.[10]

Adult boll weevils are small, about a quarter inch in length, with rounded reddish brown and gray bodies and a long curved proboscis. (See Figures 4 and 5.) Weevils have large, bulging black eyes. Spurs on the joints of its front legs are one of the few marks that distinguish it from other weevils. Due to cotton's chemical makeup, the boll weevil is attracted almost exclusively to the plant. Unlike most insects that feed on amino acids, the boll weevil requires both amino and twenty-three different fatty acids, finding their required balance of nourishment principally in cotton fibers. Weevils convert the plant's terpenoids—a chemical compound that gives off a particular odor—into pheromones, a substance given off by the weevil that attracts sexual partners. This reliance on a component of the plant to court mates for reproduction drives the boll weevil to spend its entire life cycle in and around the plant.[11]

Each year as the weather warms, boll weevils wake from a state of diapause. Depending on location, longitude, altitude, and rainfall levels, they emerge from this hibernation between early June and July. As cotton plants begin to poke through the dirt, weevils are drawn to them, feeding on their terminals and leaves, which first appear forty days after planting. These "overwintered" first-generation weevils can live for more than eleven days without food, and typically feed for three to seven days before they search out mates. After mating, females begin looking for cotton squares to house their eggs. These squares, the enclosed flowers of the cotton plant, emerge on the branches roughly fifty days after planting, depending on environmental factors. When a female locates an available square, it punctures it with its proboscis, and drops a single egg inside with its ovipositor. It then seals the hole with a frass plug, a yellow wax that protects the eggs from the outside world.[12]

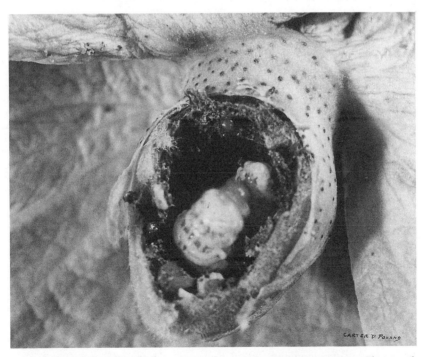

FIGURE 5. Boll weevil grub inside cotton square, magnified about 200 times. Photograph by Carter D. Poland of Anniston, Alabama. Carter D. Poland Photograph Collection, Record Group 268, Auburn University Special Collections and Archives.

The eggs then grow in this enclosed space, safe from wind, sun, and poisons applied to the outside of the plant. After about three days, the eggs hatch into larvae, which continue to eat the inside of the square or boll. The larvae live for a week or more before entering the pupae stage. While still inside the protective cocoon of the square, a massive cell restructuring follows for roughly four days, during which the pupae transform into adult weevils. This process destroys the square, which flares and falls to the ground, or hardens and dries out but remains attached to the plant. In either case, the square can no longer produce healthy fibers. When the adults are ready, they cut themselves free from the dead squares and begin their own search for food and a mate.[13]

The weevil's facility for destroying a field of cotton comes from its reproduction ability. A first-generation female that emerges from hibernation can produce only one hundred or fewer eggs during its life, though females from subsequent generations produce three hundred or more eggs. Depending on the latitude, two to seven generations of weevils will reproduce in one growing season, with each generation increasing in population tenfold. This means that weevil populations grow dramatically as the summer progresses and are at their highest when cotton is ripe and ready to be picked. The rate of reproduction is astounding. A single pair of boll weevils can theoretically account for well over *twelve million* offspring during one growing season. (The most conservative of entomologists' estimates still place the number at two million.)[14]

Weevils begin to decrease in number naturally only when their food source disappears. In late fall, after ripe cotton has been picked, the pests begin a search for hibernation sites. They will continue to eat cotton plants that are not plowed under at this stage, and if they find some nourishment this late in the season, they are more likely to survive hibernation. Boll weevils overwinter in any place where they can find protection, often in foliage near fields, in high grass along fence rows, at the edge of outbuildings, or even in Spanish moss hanging from a nearby tree. Spring wakes those insects that survived hibernation, and they emerge looking for the nearest cotton plant to begin the cycle again.[15]

The degree of injury weevils can execute on cotton differs depending on a variety of factors that are important to the history of the boll weevil, including variables in geography, topography, weather, soil type, elevation, labor practices and arrangements, cultivation methods, technologies, cotton variety, fertilizers, and countless others. The full range and complex interplay of these factors that determine the weevil's ability to destroy cotton was unknown to Tyler Townsend in 1894; he had, after all, studied the pest for only a few weeks before releasing his report. Townsend did, however, make many discoveries about the pest's feeding and life cycle that formed the basis of his recommendations for Texas farmers for checking the spread of the weevil.

Townsend submitted his report to Entomology Bureau chief Leland Howard on December 20, 1894. It is a remarkably detailed document considering Townsend's relatively short study of the weevil and what little was already known about the insect. In addition to several pages outlining everything he had gleaned about the weevil's life cycle, eating habits, and hibernation schedule, Townsend identified other key environmental factors that governed the weevil's success against cotton. For instance, he listed a few possible natural enemies of the beetle, the effect of frost dates on hibernation, and impact of rainfall on the beetle's migration.[16]

The most important section of his report contained his recommendations for Texas farmers. First, he advised those whose fields were already infested with the pest to apply either of two poisons, Paris green or London purple. For decades the USDA had recommended use of these arsenical compounds on nearly every insect pest that had appeared across the country. In addition to being expensive, the poisons tended to kill the plant along with the insect if not applied carefully. The fact that weevils hid for much of their lives inside cotton squares compounded the matter. Townsend recognized this, and only halfheartedly endorsed poisoning. He described the practice of applying insecticide to cotton once weevils had entered the safety of the square as "hopeless."[17]

The more durable of Townsend's recommendations came to be known collectively as the "cultural method." For farmers not already beset by boll weevils, Townsend suggested they perform a variety of tasks during the cultivation of the plant that would limit the weevil's damage. He advised that during the summer farmers keep vigilant watch for infected plants and if found, that they collect the fallen squares and burn them. This would obviously reduce the number of insects that would hatch later in the season. He also suggested that after harvest farmers burn the stalks and "trash" remaining in their fields; then, if the land was irrigated they should flood it. Townsend suggested farmers then turn livestock out into the fields to consume any remaining bolls left on the ground and to trample weevil hibernation spots. Finally, he advised farmers to rotate fields so that cotton was not grown on the same land year after year.[18]

In sum, these cultural methods targeted the habitat of the boll weevil; they sought to reduce the number of insects that survived hibernation through practical farming methods, rather than simply applying poison. For twenty years following the release of the report, this advice formed the kernel of the recommendations all government agents and scientists provided for southern farmers. Taken together, his suggestions were basically techniques for more careful cotton farming.

Townsend did alter his advice for one small group of farmers living inside the boll weevil territory. He argued that because the pest was still relegated to southern Texas, if farmers within a fifty-mile-wide strip just north of the Rio Grande stopped growing cotton altogether for one year, the weevil wouldn't spread. Townsend knew the economic landscape as well as the agricultural one, however, and recognized that this plan could never be accomplished with a voluntary program. Too many cotton farmers had too much at stake, so he called for "compulsory legislation" to stop farmers in the area from planting cotton in 1895. This strip of land, he noted, could support other kinds of agriculture, therefore farmers could still earn a living. Once the boll weevil spread to the north and east, into more densely planted cotton areas more dependent on that crop alone, there would be little chance for a quarantine.[19]

Assistant Secretary of Agriculture Charles W. Dabney read Townsend's report and soon boarded a train for Texas. There he quickly became convinced of the importance of a cotton ban. Dabney was a son of the South—he believed he could communicate with cotton growers and their political representatives—and he was a personal friend of the new Texas governor, Charles Culberson. He conferred with the governor and even addressed the Texas legislature to convince them of the cotton ban's merit. With the pest still relegated to a small portion of the state, however, and with no detailed plan to destroy it once cordoned off, Dabney and Townsend had a rough time convincing a Democratic governor and legislature of the need for intervention. (Interestingly, these local Texas Democrats were breaking ranks with the state's representatives in Congress, who would later argue that only the government could enforce a solution to the spreading weevil.) In the end, Dabney and Townsend's efforts failed. State legislators balked at legislating what crops constituents could grow or not grow. As the 1895 season came to a close, the legislature refused to act and the boll weevil moved on, but USDA officials had learned a lesson: lawmakers did not fully grasp the danger that the boll weevil posed.[20]

By the end of 1894 the USDA had known of the boll weevil's presence only three short months, but the bug had captured the attention of the entire department. Together, Townsend's report and Dabney's visit to the infected areas reveal how the USDA was beginning to understand the weevil's threat in two related ways. First, the department saw the insect as a critical threat to cotton and therefore to the agricultural economies of the entire South. Townsend's report made it clear that this was a pest that could devastate cotton from Texas to the Carolinas without a major government intervention. This led the department's leaders to a second, more lasting realization. The USDA wanted to stop the pest, to protect cotton from the bug's spread, but at the same time its top officials recognized that the department could never adequately fight the boll weevil without a radical increase in the department's size and influence. The federal department had a limited research budget of its own, weak connections to the investigations done on state research farms, and more importantly, no effective way of reaching the average southern farmer with any practical advice for thwarting the weevil.

From Agriculture Secretary J. Sterling Morton to field agents like Townsend, there was a recognition within the USDA that previous fights against insects had been hampered by limited legislative help and few effective ways to educate farmers. Investigations of the citrus scale in 1868, the Rocky Mountain locust in 1877, and the cattle tick in the 1880s established the precedent for federal resource allocation to regional farming problems, but these far-ranging, relatively short-term insect outbreaks led to no permanent changes to the organization of the federal farm and entomological agencies that undertook the studies.[21]

Finding the means to send scientists to study these outbreaks was only half of the problem in limiting their effects on American farming. Once entomologists like Townsend ascertained enough information about insects to make recommendations for stopping them, there was still the problem of convincing farmers to follow that advice. Congress had slowly built a network of institutions to carry out farm research and education, but in sum these efforts had had little effect on the way southern farmers actually farmed. From the Civil War to 1890, Congress passed the Morrill Act, which gave land to each state for the establishment of an agricultural college, the Hatch Act, which provided federal money for research on state farms, and the Second Morrill Act, which provided annual funding for land grant colleges. As Elizabeth Sanders has pointed out, these reform efforts were pushed by farmers and their congressional representatives, yet the work of the land grant colleges and research farms that had resulted from this legislation frustrated these same small landowners. Farmers increasingly felt detached from the work conducted at these institutions. There developed throughout the late nineteenth century a tangible and mutual disrespect between farmers and would-be agricultural reformers.[22]

These frustrations fell along two broad lines. First, there was little consistency in the ideology of "modern farming" pushed by reformers. This group itself was an odd coalition that included college professors, internationally trained agricultural scientists, rural-based politicians, newspaper editors, railroad companies, small business owners, and New South boosters. Individually, they called on farmers to modernize by both increasing their production of cotton and diversifying into other crops. To many, a modern farm was an industrialized one, geared for efficient production on a massive scale. For others, modernizing meant abandonment of the one-crop system that had held many southern farmers in a precarious state of near-poverty year after year.

These observers of farm life—indeed there were very few actual farmers among them—paid little attention to the main reason landowners planted cotton every year: they could not afford to grow anything else. The crushing weight of the South's cash shortage caused by the Civil War and perpetuated for the decades that followed tied farmers to crops that they could turn into cash at the end of the season. No farms were truly "self-sufficient." Farmers needed money for tools, fertilizer, and seed that they could not produce themselves. Cash was scarce, however. Credit trickled down from New York City, through Memphis, Atlanta, and New Orleans, then to southern merchants, banks, and eventually to landowners, most of whom passed it on to tenants. Credit tied the merchant to the farmer, and since lenders understood that cotton was the only way to make a southern farm pay in cash, they forced farmers to grow cotton again and again. These forces brought riches to many banks, some merchants, and a few planters, but bound millions of farmers to a cyclical life of credit, debt, and poverty.[23]

The second problem with the movement to modernize southern farms was in the pedagogy of rural education. This is where the distrust between farmers and reformers is seen most clearly. In the late nineteenth century, there was seemingly no effective means of teaching farmers how to employ any of the new agricultural discoveries made at the land grant schools. Researchers' findings concerning seed selection, soil nutrition, and plant pathology were worthless if yeomen farmers never learned and applied them. The principal difficulty in translating experimental results from the research farm to the average cotton field was a cultural distance between educators and farmers. Scientists saw cotton growers as simple, uneducated, and unwilling to learn. Farmers saw educators as elitist agents of an intrusive government. At the turn of the century, many college-trained researchers saw southern yeomen as historian John D. Hicks did looking back in 1931. The "ignorance of the southern farmer," Hicks wrote, "was indeed so complete that most of the propaganda for diversification, so common in the South from Granger times on, was utterly unintelligible to him, if it reached him at all, and doubtless he would have been incapable of acting on such advice even if he had known what it was all about."[24]

Conversely, farmers' perception of educators, a group that included anyone who believed science could improve the techniques learned from the collective expertise gleaned from generations of work by family and neighbors, was just as negative. One scholar has described farmers' "frequent attitude of suspicion and often outright hostility toward 'the government' run by Yankees up North." Some of this skepticism was certainly fair. Leafing through the early bulletins of the Texas Agricultural Experiment Station, for example, it is clear why no average farmer could have benefited from the information offered by the state's farm experts. The reports are filled with technical charts and graphs, and references to chemical compounds and formulas that only trained chemists or botanists comprehended. At the turn of the century, despite increased federal and state funding for research and teaching in Texas and across the South, very little practical knowledge had made its way to farms.[25]

While most federal lawmakers did not recognize the systemic problems with agricultural research and farmer education, field agents like Townsend and Knapp undoubtedly did. The case of the boll weevil outbreak in Texas made it clear that two things were standing in the way of mounting a serious campaign to limit the bug's damage. First, there could be no effective fight against the boll weevil that did not have the support and cooperation of local political leaders. Second, policy makers and business leaders must be made to realize the enormity of the boll weevil's threat. Without an understanding of how the pest could devastate their cotton-based economic and social systems, agricultural educators would not receive the necessary financial and social support they needed to fight the weevil.[26]

Two attempts to secure increased funding during 1895, one at the state and one at the federal level, elucidate these problems. In its annual report that year, the Texas Agricultural Experiment Station (TAES) clearly stated its case for more funding and manpower. Professor J. H. Connell, Texas's state horticulturalist, recalled that USDA Assistant Secretary Dabney, on his earlier visit to Texas, had suggested that Connell himself lead the state's organized fight against the boll weevil. In his report Connell firmly rejected the suggestion. He was already the sole investigator in both the horticulture and entomology divisions of the station, he pointed out, in addition to his teaching responsibilities at Texas A&M. "Consequently," he wrote, "no more additional work could possibly be undertaken without cutting down the work more or increasing the force."[27]

Connell's comments suggest the patchwork nature of farm research and education at the time. Federal law limited USDA intervention into the boll weevil outbreak to local investigations and support of state-level research. Connell clearly wanted more staff and money to mount a serious attempt to educate farmers on methods to limit the weevil's effect, but was unable to do so. Even the vehicle through which he made this point—the end-of-year report for the TAES—suggests how far removed from actual funding debates his department was, and more importantly, it contextualizes the way that his state department would use the boll weevil to move these discussion into a more public arena.

The second attempt to gain funding for the boll weevil fight in 1895 happened at the federal level. Early that year, an unlikely voice in favor of government intervention emerged on the floor of the U.S. Senate. On February 17, Newton C. Blanchard waved a collection of papers in front of his fellow lawmakers. "I hold in my hand certain letters," the Louisiana senator told his colleagues, "which called to my attention . . . a new enemy of the cotton plant that has appeared during the past season in southern and western Texas." Blanchard's pronouncement was the first public mention of the boll weevil in Congress. The Senator asked his colleagues for a $100,000 appropriation "for investigating the spread and devastation of the cotton-boll weevil . . . and for experiments looking to its destruction and eradication."[28]

With a mix of foresight and rhetorical savvy, the senator warned his colleagues that "if unarrested the weevil will rapidly increase and extend its ravages in new directions, so that in time the entire cotton belt will be . . . affected." Blanchard portrayed the boll weevil as more than just a localized scourge; it was a foreign invader that warranted a federal response. "The *Mexican* weevil, while as yet confined to a portion of Texas, is actually a menace to the entire cotton belt," Blanchard warned, arguing that because of its possible spread to other cotton states "it is considered that the matter is a fit subject for national legislation." His Senate colleagues were unmoved. At the close of the 54th Congress, Blanchard's bill was dead.[29]

A closer examination of the Louisiana senator's attempt to garner federal funding for the boll weevil fight suggests exactly which southern constituents were most concerned about the advancing nemesis. As Blanchard spoke, the boll weevil was still over 125 miles from the border of his home state, having only moved roughly that same distance in four years. Despite the rhetoric, his concern was not for the cotton growers of his home state, but rather for the million-dollar cotton marketing industry of New Orleans. The letters Blanchard waved above his head on the Senate floor were not from desperate cotton-raising constituents who feared the advance of this killer into their fields, but rather these were letters from planters writing to the president of the New Orleans Cotton Exchange concerned about the weevil's effect on cotton pricing. While Elizabeth Sanders has convincingly documented the political power of farmers to lobby state institutions in this period, evidence suggests that in the first few years of the weevil's presence in Texas they did little to directly impress upon political leaders their concern about the pest.[30]

*

Both state and federal efforts to fund more research into the boll weevil in 1895 failed, and the repercussions of this inaction became increasingly clear as the weevil destroyed more and more cotton. In 1895, reports came from farming communities on the southern border claiming the insect had destroyed the entire crop that year.[31] By planting time in 1896, the weevil was as far north as San Antonio, and Texas's farmers still had little understanding of the weevil or of how to manage it. Over the next few seasons the pest continued to spread. By 1901, the weevil was in Waco and Palestine; two years later it was as far east as Nacogdoches, all the way to the Red River in the north, and past Dallas to the west. (See figures 1 and 2 in Introduction.) By 1903, when the weevil first entered Louisiana, it had infested five million of Texas's seven million acres of cotton land. The resulting destruction was impressive. One year the weevil had destroyed enough cotton to account for 300,000 bales, an amount worth close to $15 million.[32] Though Texans continued to buy land to the west of the weevil and plant more cotton there, farmers within the insect's grasp suffered 25 to 75 percent crop losses.[33]

In light of the pest's spread and continued impact on the state's key crop, local educators, state scientists, and cotton farmers all sought ways to get Townsend's cultural recommendations for fighting the weevil, as well as any of the findings of researchers at the state's experiment farms, into the hands of those who could enact them. Since 1894, state and federal officials had told anyone and everyone who would listen of the dangers posed by the weevil, but simple warnings about future crop devastation had proven ineffective. What they needed was a way to speak to farmers directly, a method that would, with little governmental aid, allow reformers to show landowners firsthand the benefits of following their recommendations.

A new educational system — one tailored for cash-strapped government agencies and weevil-plagued farmers alike — arrived in the form of Seaman Asahel Knapp, a "grizzled Victorian of seventy with white muttonchop whiskers, piercing eyes, and indomitable energy," as C. Vann Woodward described him. There are few figures in agricultural history more revered than Knapp. Scholars have attributed nothing short of a farming revolution to this "Schoolmaster of American Agriculture," our "greatest agricultural statesmen," though his popular and scholarly image has been distorted by the context in which he rose to fame: the boll weevil's spread through Texas and Louisiana. Legend would have it that Knapp all but singlehandedly beat the boll weevil in Texas and devised there a unique and effective program for convincing farmers to engage in modern, scientific methods. In truth, he implemented in Texas an already proven system for farmer education by hyping the boll weevil threat and his own solution for it.[34]

Knapp's work did in fact revolutionize the relationship that southern farmers had with their government, but his role as a spokesman for the boll weevil fight was almost as lasting and important, as evidenced in how he has been remembered. To one observer singing his praises in the late 1920s, Knapp was a "venerable but dynamic seer [who] threw back the invasion of the cotton boll weevil." To another, his fight against the pest was the "greatest single piece of constructive educational work in this *or in any* age." "No other two men" than Knapp and Booker T. Washington, wrote another, "have done more for the Negro in the lower South since Emancipation." Perhaps the greatest praise was heaped on Knapp by Jackson Davis, an extension worker in Virginia, who claimed "There was a man sent from God, whose name was Seaman A. Knapp."[35]

Knapp's canonization began with his work to help farmers fight the boll weevil in Texas and Louisiana. He helped to create there a new system of farmer education, but his real genius was in his ability to reframe the meaning of the boll weevil itself, from a local nuisance to an unconquerable foe, a truly national natural disaster. Knapp was as much a public relations expert — a showman even — as he was an educator, and by making the boll weevil into a menace worthy of the nation's attention he in turn remade the American farmer's relationship with the state. His work promoting the boll weevil myth was far from his first chance to put his promotional skills to work.

Before becoming the celebrated hero of planters, politicians, and businessmen, Seaman Knapp was a carpetbagger. Born in rural New York in 1833, Knapp moved to Iowa, where he raised sheep, tried his hand as a Methodist minister and a school superintendent, and eventually became a professor of agriculture at Iowa State Agricultural College in 1879. From 1883 to 1884, Knapp even served as the school's president. The Hawkeye State had been the first to build a Morrill Act land grant college, which had opened its doors in Ames only a decade before hiring Knapp. In addition to teaching husbandry, Knapp became a leader of the state cattle breeding association and editor of the *Western Stock Journal and Farmer*.[36]

After five years in Ames, Knapp left in search of a more lucrative position in the private sector. Alongside hundreds of other educated northerners, he moved south in search of a way to transfer his talent as a farm expert into private business. The land speculation prevalent in Texas when the boll weevil crossed the border took place simultaneously in other "underdeveloped" parts of the South. Railroads pushed into the hinterlands of Louisiana, as they had Texas, and midwestern settlers briskly gobbled up land there. The North American Land and Timber Company had acquired nearly 1.5 million acres of soggy Louisiana marsh, which most farmers did not consider arable, and the company was looking for someone with rural credentials to convince immigrant farmers that they could in fact make a crop on the wet land. A company representative who had known Knapp in Ames asked him to serve as its assistant manager. He agreed, and moved his family to Lake Charles. Knapp quickly involved himself not only in farmer relations, but also in mortgage operations and a host of endeavors independent of the firm. He was soon knee-deep in timberland speculation, as well as the supervision of twelve large rice farms, sawmills, and sugar operations.[37]

Though Knapp's official concern was making the soggy land profitable for his company, from 1891 to 1903 he relied mostly on his talents as a promoter. He became the leading advocate of rice culture in southwestern Louisiana and a major player in the agricultural transformation of the region. Not only did he convince farmers to buy the wetlands and to grow rice, but he also pushed for the use of new technologies, including mechanical seed planters, harvesters, threshers, and pumps that could raise and lower the water level of rice paddies. Knapp parlayed his interest in rice production into a vast business. Along with several partners, he built rice mills, founded a Lake Charles bank, organized the Rice Association of America, and created and edited *The Rice Journal and Gulf Coast Farmer*. His focus was seldom on improving the lot of the small Louisiana rice farmer, however. The core of his work was as a booster for rice agriculture and the Louisiana businesses that profited from it. As a result, farmers themselves rarely heralded Knapp or his accomplishments nearly as much as railroads, merchants, banks, and politicians.[38]

In 1902, USDA officials recognized Knapp's ability to move seamlessly between farmers and the business sector and contacted the former educator about a job promoting southern agriculture. The agency needed a representative to convince cotton farmers to follow Townsend's cultural recommendations and it hoped Knapp could help. The old professor, already seventy years old, accepted the department's challenge and began working to establish farms on which to demonstrate the cultural method to skeptical farmers.[39]

Knapp's initial goal was to confront these farmers with convincing demonstrations through a system that he had used with rice growers in Louisiana. This idea of modeling farm work for local landowners was nothing new. For decades, states, as well as railroads interesting in shipping more farm products, had asked farmers to visit research farms. But for most landowners this meant a long trip to visit a demonstration farm. The alternative was for educators to simply mail printed bulletins to rural people, or to have the information reprinted in newspapers. Knapp was not the first reformer to identify the problems with this system, but he was the pioneer in using the boll weevil to find a solution.[40]

In the late nineteenth century state-employed botanists, chemists, and other researchers had to physically take their recommendations out to farmers by establishing local, private demonstration farms. The problem, reform-minded educators like Knapp understood, was not the demonstrations themselves, but the limited number of them. It was too costly to set up a multitude of local demonstrations because in each case the state had to guarantee the landowner against loss. (A local farmer would agree to hand over part of his land for the demonstration only with a promise from the state that he would not lose any money.) More demonstration farms would mean that men and women would not have to travel to remote government plots to see the latest fertilizers, newest crops, or latest tools, but Knapp recognized that the costs were too high.

At the start of Knapp's work as USDA special agent, there was no set demonstration agreement between the government and volunteer demonstrators. Agents made a variety of arrangements with local farmers to establish demonstrations. In most cases, landowners gave up a portion of land, their buildings, and their tools to the USDA for one year; in return, the department paid the costs for labor and supplies (seed, fertilizers, pesticides, etc.). At the end of the season, the landowner and the USDA divided the crop equally. In some cases the farmers were guaranteed a profit based on the average yield of the surrounding county. Though this technique was successful in convincing individual landowners of the benefits of modern farming practices, it failed to exert much influence on neighboring farmers, which was the point of the demonstration. Effectively teaching a large number of farmers with a demonstration of this type still meant that people had to travel to visit and observe the USDA's methods on a nearby farm; most farmers were unwilling to take time away from their own land to visit a model farm.

Even for those who would visit the farms, adopting the department's recommendations was risky. Selecting a new kind of seed, rotating land in a new fashion, or even planting new crops was an expensive gamble, and those whose profits were not guaranteed by the USDA could not be made to place that bet. In other words, even if a farmer who visited a demonstration farm was swayed by the USDA's example there, he was unlikely to put what he learned into practice on his own land.

Another reason for demonstration farms' ineffectiveness was that much of the soil and seed improvements Knapp prescribed took several seasons to show their benefit, though arrangements with demonstrators were usually limited to a single season. Knapp tried to convince farmers to sign on for several years, but most refused. Since the demonstration technique could not be effective under these short lease agreements, the results after a single year were seldom significant enough to persuade farmers that the government's methods were superior. Most farmers remained uninterested.

Despite Knapp's initial failure to convince many farmers to adopt the USDA's cultural recommendations for limiting boll weevil damage, southern newspapers publicized his work, and it caught the attention of merchants, railroad executives, and other businesspeople. Most recognized that as the boll weevil proceeded into new territory, all business in cotton country was threatened. They also recognized that local farmers would need to be persuaded to follow to the department's advice. When Knapp's work became better known, local businesspeople across Texas and Louisiana contacted the USDA, asking to have Knapp, or Beverly Galloway, the head of the USDA's Plant Industry Bureau, establish a demonstration farm in their town.[41]

In 1903, a group from Terrell, Texas, contacted Knapp about setting up a demonstration. Because Terrell was a fair distance from the Louisiana demonstrators Knapp was already committed to working with, he feared spreading his operations too thin over a large area and declined Terrell's offer. Knapp received a second letter from Terrell a few weeks later, however, this one from an agent of the Texas Midland Railroad. Terrell was the home of Edward H. R. Green, owner of the rail line, and a famously wealthy Texan. (This son of a railroad tycoon father and an independently wealthy mother, Green had apparently arrived in Terrell unannounced in 1882, walked into the local bank and deposited a quarter-million dollars in cash. By the time he left the bank that morning he had been placed on the bank's board and made a Texas "Colonel.") Green purchased land in the area and developed and managed the Texas Midland Railroad, which quickly became an important link between the increasingly dense cotton hinterlands and the state's shipping centers.[42]

The railroad's letter to Knapp set it itself apart from others requests not only because of the famous connections of the line's owner. The railroad agent assured Knapp that a local group, calling itself the Terrell Farmer's Institute, was willing to raise money to guarantee a local demonstrator against loss. This was a new twist in the demonstration arrangement, which had traditionally relied on USDA funds to guarantee demonstrators. Knapp had been reluctant to accept Terrell's invitation because of his commitments in Louisiana, but the promise of the Farmer's Institute covering the bulk of the cost of the farm made the proposal worth looking into.[43]

Galloway and Knapp traveled to the east Texas town and found it in what Knapp would later identify as a state of panic. The boll weevil was due in Kaufman County the following year, and farmers had heard horrific tales of the devastation it caused in the cotton fields to the south. The educators agreed to talk to a meeting of concerned citizens and to examine the prospects of a demonstration farm in Terrell. On February 25, 1903, they arrived at the town's Odd Fellows Hall for their first meeting, surprised to find the room overflowing with people. As Knapp stepped to the dais, a farmer shouted to him from the crowd, asking what solutions for the weevil he had brought them. "Not a thing," Knapp reportedly countered. He did not have a magic solution to stop the insect, he explained, but he did have a plan of self-help to improve their farms. "Any scheme of relief that is not based on self help is like sending a man to hold up a sick calf," he told the group, "after a while they both get tired and fall down together." He lectured the audience on a broad range of methods they could employ to increase productivity on their farms: careful selection of improved seed, soil rotation, improvement of underused lands, and a steady application of approved fertilizers. Surprisingly, he made no specific mention of a course of action to thwart the boll weevil.[44]

After the meeting, Knapp pressed the leaders of the Farmer's Institute on their promise to fund the demonstration farm. Within a half-hour the group had collected $415 to guarantee a demonstrator against loss. Impressed, Knapp agreed to stay overnight and to visit potential demonstration farms the following morning. He must have been encouraged by the interest shown in Terrell, even though it was principally expressed by business leaders, not farmers. Knapp must also have realized that the impetus for the demonstration seemed to spring not from a concern for farm improvements, but from unmitigated fear of the boll weevil. Whatever the reason, Knapp recognized that with a cash outlay from townspeople and a motivation to learn, the atmosphere in Terrell stood in stark contrast to the flagging demonstration spirit in Louisiana.[45]

The next morning, Knapp and members of the Farmer's Institute visited several landowners who had volunteered to serve as demonstrators. Eventually the group selected the farm of Walter Porter, despite the fact that it was three miles from Terrell, which Knapp considered a drawback because farmers and townspeople would most likely not travel on foot to observe the farm's progress. Knapp established an executive committee consisting of farmers and business owners to oversee and implement his recommendations for Porter's demonstration. On February 25, 1903, the agreement was formalized and signed. Porter, the committee agreed, could keep any and all profits he made on the farm and the committee would compensate Porter at the end of the season if he lost money implementing Knapp's advice. The crucial difference between the Terrell farm and Knapp's previous demonstration efforts was this guarantee by the local business community against any loss suffered by Porter. The USDA was not required to put up any initial outlay of money. Perhaps more importantly, it assured local farmers that the town's business community was a stakeholder in the weevil fight.[46]

Per Knapp's instructions, Porter devoted seventy acres to the demonstration, planting thirty-seven in cotton, twenty-four in corn, and the remainder in a variety of vegetables and ground cover. The cotton was divided into nine plots, on which Porter systematically experimented with a variety of seeds, fertilizers, and soil types. The corn crop took a beating from an unusually wet season; wind and rain reduced the yield more than 50 percent. But the cotton flourished. As expected, the insect did march into Kaufman County that year; however, it never made it to Porter's cotton fields. Despite the hysteria surrounding its invasion of nearby land, the pest never bothered Porter's cotton in 1904. At the end of the season, while many neighboring farmers directly to the west and south had lost huge portions of their crop to the weevil, Porter revealed that his experimental cotton had been wildly successful. One plot yielded over 325 pounds of cotton lint per acre, nearly twice the amount from traditional farms. His experimental plots earned Porter $700 more than his other land. The combination of a bumper cotton crop and the *assumption* that it was made in the presence of the boll weevil combined to create the immediate and widespread legend that Knapp and Porter had beaten the boll weevil with this demonstration method.[47]

The influence of this single experiment farm cannot be overstated. Despite the reality that the farm made its profits under relatively good conditions, not while the weevil was present, across the South people latched onto the notion not only that a boll weevil remedy had been discovered, but that out on Walter Porter's seventy acres federal agents had hatched a new system for teaching rural farmers how to pull themselves out of a cycle of poverty. Knapp had been using this same basic demonstration method for months, but the Terrell farm was the first to be considered (however erroneously) to be located within the boll weevil's territory, and thus because of Knapp's ability to publicly frame the weevil as a profound threat to the South's economic future, the success of the Porter Farm became national news.

Word of Porter's perceived success spread faster than the weevil itself. Southern newspapers broadcast news of the farm all over the region. Knapp, for his part, encouraged the perception that the farm had been a wild success; he knew that if rural southerners believed he held the key to profitable farming in the face of the boll weevil, they would listen to whatever it was he had to tell them. Consequently, Knapp began receiving countless invitations for speaking engagements all over the South, especially from areas that lay in the boll weevil's immediate path. Like Knapp, USDA officials took advantage of the public relations victory offered by the Terrell experiment. At the end of the 1904 season, Secretary of Agriculture Wilson traveled to Terrell to discuss with Knapp the means of applying this new demonstration method across the South. Wilson and Knapp agreed that this arrangement was ideal because it required no outlay of funding from the USDA, only a guarantee against loss raised by the local business community. The government therefore only had to fund agents' salaries and transportation costs. Wilson immediately went to work on southern politicians to take advantage of the Porter Farm's celebrity and to lobby Congress to appropriate funds to spread this demonstration method across the South.[48]

Proponents of a federal demonstration bill used the harrowing reports of boll weevil damage in Texas to get lawmakers' attention, then promised them that a new, effective method to teach farmers how to fight the pest was available. Luckily for Wilson and Knapp, the boll weevil served as a powerful lever to spur congressional action. In December 1903, only weeks after Porter's demonstration cotton had been harvested, the U.S. House Committee on Agriculture considered a bill to specifically fund the USDA's boll weevil fight. House representatives from Texas testified as to the disaster that the pest had created on the state's cotton farms. George Burgess pointed to a 50 percent crop loss in his home county of Gonzales for the 1902 season.

Democrat Scott Field of Robertson County claimed tenants on his plantation had picked 1,700 bales in 1901 but managed a paltry 103 bales the following year. USDA officials echoed these complaints, reporting that the boll weevil had accounted for a $15 million dollar loss in cotton production from the previous year. Leland Howard, Entomology Bureau Chief, also testified before the committee, and brought with him a two-foot long papier-mâché model of a boll weevil. When called to testify, Howard unloaded the gigantic bug from a dry goods box and placed it on the table. He later recalled, "Captain Lamb of Virginia turned with amazement to Congressman Burleson of Texas . . . and said, 'My God, Burleson, is it as big as that?'" Fear that the boll weevil was taking over southern agriculture was a convincing force.[49]

As the debate on a boll weevil bill continued, however, many nonsoutherners raised doubts as to whether the weevil warranted federal legislation. The irony that recent proponents of southern states' rights were now suddenly calling on the government for help was not lost on several northern Republicans. When Iowa Republican Gilbert Haugen pointed out the southerners' ideological contradiction, Congressman Field admitted, "the time was when a southern member would hesitate to go to the Government asking relief, even though the damage was exceedingly great." He seemed to brush off a century of southern political ideology by stating plainly, "times have changed, and we have modified our views." Nonsoutherners on the committee stood fast in their claim that the boll weevil did not represent a threat to the nation and that any agricultural legislation that affected one part of the country should affect the rest equally. As a result, northern representatives modified the bill to include funding for the investigation and eradication of foot and mouth disease, which primarily afflicted the dairy cattle of the Northeast and Midwest. In late December, the committee passed the bill, and it moved to the House floor.[50]

As in committee, the House debate over the weevil bill fell mostly along regional and political fault lines. Proponents again used the fear of the boll weevil's spread over the entire Cotton Belt to attract support. "This is a bill of great importance, not alone to Texas, but to all States engaged in the production of cotton," Representative Burleson said to begin the House debate. Painstakingly describing the boll weevil's assault on Texas cotton, then explaining the speed at which the weevil was traveling towards the rest of the cotton belt, Burgess attempted to reduce the perception of the boll weevil as merely a regional problem. But many from outside the Cotton Belt were not swayed.[51]

The objections of Frederick H. Gillett, a Massachusetts Republican, typify the debate made by nonsoutherners in opposition to the boll weevil bill. Gillett downplayed the seriousness of the cotton insect's national importance by offering a lengthy and detailed discussion of the "extraordinary history" of the gypsy moth and the "national evil" of its infestation of Massachusetts' trees. He suggested federal legislation to deal with the moth, but his larger rhetorical point was that Massachusetts had been fighting the moth without federal help. This northern pest, Gillett argued, was more deserving of federal attention than the boll weevil because it attacked "everything that is green," unlike "this specialist from Mexico" that attacks only one cotton.[52]

The debate over funding to fight the weevil exemplifies not only the legacy of sectional division within the federal legislature, it underscores the failure of many outside the South to comprehend just how complete a threat the pest was to southern society. To southerners and their federal representatives, this natural enemy to cotton had become a threat to the very social and economic fabric of the region. Absent from the debate on either side was the issue was the South's unwavering commitment to cotton itself. No one rose up in opposition to the bill and suggested that the South grow less of the crop. Also curiously absent from debate on the bill was any talk of the North's own reliance on the health of southern cotton. Mills in the northeast were still the principal consumers of the southern staple, yet southern proponents failed to make that point. Nor did northerners seem too concerned about the supply of cotton to these local industries.

On January 13, 1904, the Senate passed a bill appropriating $250,000 to the USDA to fight the boll weevil; the bill contained a clause directing any excess funds to the eradication of foot and mouth disease in cattle. President Roosevelt signed the bill two days later and the funds were made immediately available for the extension of Knapp's system and the hiring of new federal agents to implement the Porter Farm model throughout the boll weevil territory. The Farmers' Cooperative Demonstration Work was born.[53]

With the news of Congress's action Knapp immediately established a base of operations in Houston and began hiring agents to launch demonstration farms across Texas. By the end of 1904 he had hired twenty men. Knapp assigned to each new hire a territory within the infested region, mostly in Texas but including some land where the weevil was present in Louisiana and Arkansas, and charged him with establishing at least one demonstration farm in each county within boll weevil territory.[54]

Almost overnight, young white men with college educations and federal paychecks boarded trains bound for the hinterlands to press the flesh in rural stations and to knock on farmhouse doors. Agents offered help to the most prominent farmers first, hoping to secure their cooperation and to establish a demonstration farm on their land. Most of their early efforts were geared toward these demonstrators, trying to make sure that each volunteer followed the department's advice precisely. People who visited the demonstration farms and promised to put the government's recommendations to work on their own land were considered "cooperators," but agents rarely had the time to visit or aid these men and women directly. In addition to the demonstration work, agents spoke to crowds assembled in churches and halls, on farms and in the courthouse. Anyone was a potential audience for their message of beating the boll weevil with early planting, selective seed use, and reducing cotton acreage.[55]

Following passage of this federal boll weevil bill in 1904, two important changes were in process. First, the impression that Knapp had slain the weevil, a pest more imagined than real on the Porter Farm, had convinced Congress to create a network of local demonstration farms. Fear over the boll weevil had fundamentally shifted the way agricultural knowledge was to be disseminated. Agents now had a federal mandate to take agricultural education to the fields. Farmers who had never traveled to state experiment farms or attended lectures by farm experts could now find advice in their own counties, at least for those who lived within the boll weevil territory. The teaching method associated with Knapp fundamentally changed how farm knowledge could move to the rural areas. The boll weevil threat, lawmakers and farm reformers hoped, would finally convince southern cotton farmers to listen to what the government had to say.

The second shift that coincided with the creation of the demonstration system was the increase in the boll weevil's notoriety. Agents intentionally heightened the drama of the invasion to use it as a tool to reach rural landowners. These educated, motivated representatives of the USDA were determined to make the boll weevil a cause célèbre. With enough public attention focused on this threat to the South's agricultural economy, these men and women hoped to reinvent and reinvigorate farmer education. Not only would state governments reengineer their relationship with federal authority, as money, people, and resources flooded into weevil-infested areas, but individual farmers' relationship with the state would be forever changed as well. The boll weevil held the key, demonstrators and other reformers believed, to continued financial and institutional support for the improvement of southern agriculture. Whether or not fear of the boll weevil and this new system of farmer education would actually change the way southern landowners and their tenant labor farmed the land, however, remained to be seen.

Cultures of Resistance in Texas and Louisiana: Tenants Make Sense of the Boll Weevil

Few if any of the South's half-million tenant farmers noticed Congressional passage of the 1903 boll weevil bill. They were too busy making arrangements for the next season. Winter was a time for decision-making, and the arrival of the boll weevil—or its impending arrival—factored into tenants' calculations. The fact that USDA-sanctioned demonstrators would be nearby had little bearing on these farmers' choice of where to work and who to work for. In fact, tenants' disinterest in Knapp's programs was mutual.

Though the success of the South's economy rested on the willing and co-erced labor of men and women on cotton farms that they did not own, tenants were, at least on the surface, absent from the early twentieth century public debate over both this legislation and the boll weevil generally. As state legislatures and Congress wrestled with policies to aid landowners fighting the pest, policy makers and government agents who controlled the weevil war rarely addressed, literally or figuratively, the people who would actually conduct the boll weevil fight. When the experts and lawmakers did mention labor, it was simply to deride it, to name it as yet another hindrance in the battle to thwart the advancing bug.

No piece of evidence demonstrates farm workers' absence from the public boll weevil fight more than a black and white photograph hidden away in a 1929 pamphlet. That year, Texas A&M published the speeches from a celebration of the twenty-fifth anniversary of the Porter experiment farm. Slipped into the middle of the slim book is a photograph that communicates more than the sum of all the words in the collection (Figure 6).[1]

FIGURE 6. Walter Porter (right) and John McQuinney (left). The caption in the *Silver Anniversary Collection* reads "Walter C. Porter and the old negro helper who worked the 100 acre field that constituted the first demonstration and which averaged between $7 and $8 per acre more than the community average that year." Negative 33-S-17943, Ed C. Hunton, copy negative, 1952, Federal Extension Service, USDA, National Archives. Photo courtesy of Texas A&M University Library.

In the foreground of the image is Walter Porter, operator of Knapp's first demonstration farm. He stands with his hands on his hips in a clean white shirt and Stetson hat. To Porter's right but a few steps behind him is a slightly older African American man wearing overalls. While Porter has a detached but confident look, the black man behind him looks tentative, even awkward. There is a dog at his feet and another behind him. The caption under the picture says only "Walter C. Porter and the old negro helper who worked the 100 acre field that constituted the first demonstration and which averaged between $7 and $8 per acre more than the community average that year."[2]

In all of the ink spilled on the Porter Farm, this is one of only two known historical references to Porter's "old negro helper." Though contemporary chroniclers and historians have written at great length, and mistakenly, of Porter's farm as the first successful battleground in the boll weevil war, unraveling the mystery of this black man's identity, let alone recording the work that he did actually implementing Knapp's advice, went undone for almost a hundred years. One could read all the extant histories written between 1903 and 2000 and not find a single mention of the idea that Porter himself did not implement Knapp's advice on his demonstration plot. The worker's absence from this story, from both contemporary discussions of the farm and historical analyses of it, speaks volumes about the way farmers at the time and scholars since have understood farm work—and those who performed it—as subordinate to expertise. Historians have eulogized Knapp and Porter, but ignored the man doing the actual work.

Historian Debra Reid, working in the 1990s on a dissertation about African American farm life in Texas, was the first scholar to ascertain the true identity of the farmer in the background. His name is John McQuinney. As Reid explains, the McQuinney and Porter families farmed together for generations, in an intertwined relationship that spanned not only decades but geography as well; the families had moved together from Louisiana to Texas in search of better cotton land. Walter Porter's son recalled to Reid how his family and the McQuinneys were tightly bound. The children grew up and played together, but as Reid points out, once adulthood set in, the relationship between white and black, between landholder and renter, asserted itself and the conditions became starkly unequal.[3]

This selective memory of Knapp's farm demonstration illuminates more than just the racial status quo in rural Texas at the dawn of the twentieth century. It is not simply that whites did not respect the work that McQuinney performed. Instead, his disappearance from public memory illuminates two problems. First, it points to a fundamental contradiction that lies at the heart of Knapp's educational movement. Knapp had come upon an effective means of reaching farmers with new ways of selecting, growing, and harvesting crops. The importance of this system lay in its ability to directly reach the person who actually labored in the fields. Yet, as John McQuinney's anonymity demonstrates, who that person was turned out to be unimportant to Knapp, his colleagues, or to generations of historians who have written glowingly of Knapp's work.[4]

Second, McQuinney's absence from historical scholarship points to a more basic problem with historical sources, one that encompasses both the case of the Porter demonstration farm and the larger history of the boll weevil. Historians do not have ready access to the thoughts, motivations, or actions of John McQuinney — they didn't even have his name until recently — or of the overwhelming majority of southern tenant farmers. The few tenants whose lives have been recorded in detail, as in Ned Cobb's *All God's Dangers*, have been overcited and everemphasized by scholars. Scholars must find more creative ways to uncover their stories.[5]

To understand the ways that landless workers grappled with the arrival of the boll weevil, this chapter explores the tenant perspective in a couple different ways. First, I offer a demographic snapshot of tenancy in Texas and Louisiana at the time of the boll weevil's initial spread to paint a picture of how the insect affected tenants' lives. Statistics can't tell the whole story, however. Tenants' own observations, recorded by landowners and social scientists, offer evidence of how the workers interpreted the arrival of the boll weevil and in turn used the pest to shape their own futures. Likewise, in the same set of sources, evidence of landowners' behaviors toward tenants, including the extent to which they worked to keep workers, is also convincing. The third set of sources sprang forth from the pressure exerted by the boll weevil onto this landowner-laborer relationship: songs and stories about the pest created by these workers themselves. Just as USDA and state officials attempted to manipulate news of the weevil to garner more support for their projects, tenants employed the boll weevil myth to carve out a better life in the rural South.[6]

*

From 1892 to 1908, as the boll weevil made its initial move from the Mexican border toward the Mississippi River, southerners had many kinds of conversations about the little beetle. There was the public one, held in Austin, Baton Rouge, and Washington, by government officials and their trusted public servants. Another discussion, carried out by the region's tenant farmers in dusty cotton fields, at crowded churches, on rickety porches, or even at raucous parties, carried news of the pest and attempted to make sense of the bug's movements and threats. These men and women made observations and became active participants in trying to shape how the weevil, along with the new system of farm education that had come along with it, would change their own economic and social status in the rural South.

No one knows what John McQuinney's agreement with Walter Porter was in 1903, but if it was like that of most black sharecroppers working on a white man's land, the arrangement was based not only on agricultural and economic conditions, but also on the local racial power structure as well. After the Civil War, farm laborers without their own land, a group that included former slaves, together with landowners in need of hands to grow crops, reinvented the system of farm tenancy. Tenant farmers included all kinds of non-landowning farmers, those who rented land outright, or were remunerated for their work with cash or, more likely, a portion of the crop. The latter form is sharecropping, which became the system of choice for both workers and landowners in the cotton South for much of the late nineteenth and early twentieth centuries.[7]

Sharecropping was not a uniform arrangement. Depending on what resources the laborer brought into the agreement—a mule, plow, or other tools, for instance—he or she would be paid a greater portion of the cotton harvested. Payment occurred only once, at the end of the season when the cotton had been picked. During the season, most sharecroppers were extended a line of credit from a nearby country store or a plantation outfitter, where the worker could buy some household items, food, and clothing. Though the majority of southern tenant farmers worked under this basic definition of sharecropping, there were a host of factors that determined their success as farmers, their freedom to move or to arrange new contracts, and their general quality of life.

With white supremacy undergirding the legal and social ways of life in the rural New South, it comes as no surprise that it also played a determining role in the system of cotton production as well. And while the common understanding of racism's role in sharecropping concentrates on whites' abuses of the end-of-season settlement, white supremacy played a role in the owner-tenant relationship year round. Observations of who worked where and under what conditions bear this out.

In 1900, 60 percent of farmers working in the densest cotton regions of eastern Texas were tenants. That figure rose slowly but steadily over the next thirty years, as cotton acreage grew. Much of this labor was Mexican. From

1890 to 1910, the number of Texas residents who were of Mexican descent doubled, the majority of whom were landless farmers. Rates of tenancy for black Texans were equally high. At the turn of the century, 69 percent of African American farmers were tenants. While only 31 percent of black farmers in Texas owned their own land, about half of the state's white farmers owned the soil on which they worked.[8]

Not surprisingly, the wages earned by these landless farmers changed as the region became more directly tied to the cotton economy, but they were consistently higher for white Texans than for nonwhites. The more the state's cotton acreage grew, which it did throughout this period, the further the average farm wage fell. When broken down by race (as complicated, inconsistent and, in hindsight, ridiculous, as the USDA's methods were) the picture is even bleaker. In 1900, black Texans earned, on average, three dollars less per month than their white counterparts—about 75 percent of the white wage.[9]

In the fields of Texas, these wage statistics meant little to the landowners and tenants bargaining for an exchange of labor for land. In fact, the relationship between owners and workers was more complicated and malleable than the statistical picture can provide. In late-nineteenth century Texas, unlike in any other region of the South that the weevil would visit, tenancy meant a diverse mix of laborers striking new deals with landholders. Sociologist Paul S. Taylor understated his observation that the "deep-seated cleavages of race and class" in southeast Texas were "often curiously aligned." Historian Neil Foley put it more pointedly:

> Throughout the last half of the nineteenth century, black/white east Texas and Mexican south Texas were converging on each other, as Mexicans gradually moved north and southern whites and blacks pushed west on the rich prairies of central Texas.

The result, Foley argues, was a clash of race and culture unique to this time and place. But this collision of African Americans, Anglos, and Mexicans in Texas coincided with the entry of the boll weevil, which disrupted this already disordered situation.[10]

Within this conflict of people, cotton itself helped determine an individual's role. In other words, how a person related to the production of cotton— whether they were in a position of "domination or subordination" in Foley's words—could help to decide the construction of racial categories. Some white landowners intentionally hired only Mexican tenants because, the owners believed, they worked harder for less pay. One grower told a researcher at the turn of the century, "the Mexicans are the only class of labor we can handle. The others won't do this work; the white pickers want [window] screens and ice-water. To white pickers I say, 'If you will accept the houses we have for the Mexicans, you can work.'" White workers' demands often left them farther and farther from the richest lands. As a result, the levels of tenancy among Mexicans soared as cotton acreage in Texas grew. In 1850, for instance, prior to the cotton boom, the Mexican rural population in southern Texas was nearly equally divided between landowners, skilled workers, and unskilled workers, but by 1900, landowners were only 16 percent of the Mexican-American population. As a result, race and labor determined each other, but were also constantly shifting according to the latest agricultural and economic conditions, which would be directly threatened by this cotton-eating beetle.[11]

Tenant farmers in Texas and Louisiana first encountered the boll weevil in conversation, not in person. In talks with other farmers, by reading newspapers, by chatting with townspeople during trips to the store, or even from the pulpit (one Texas preacher promised his congregation in 1898 that the insect was "the evil spirit that dwelleth amongst us"), they encountered first the idea of the weevil. They heard that this bug was destroying the state's principal crop, that it was a force that promised to wreck the social and economic system ordered by cotton.[12]

At this early stage of the weevil's presence in the United States, state and federal scientists and demonstration agents were the most common groups spreading this image of the weevil. By 1903, the USDA reported that the beetle had destroyed 53 percent of the cotton in places it had invaded. (It also reported that nonweevil counties had increased their production of cotton by 11 percent, at least in part to make up for the fiber consumed by the weevil.) While these statistics originated in government reports, local newspapers eagerly reprinted the dire information, carrying news of the bug's effect on Texas cotton all over the South. Knapp's army of demonstration agents also began appearing in weevil counties that year, and in an effort to drum up local support, made claims about the pest's disastrous effects on the state's farms.[13]

Demonstration agents proclaimed both the weevil's threat and trumpeted what they presented as their *unique* ability to provide solutions to the pest. Farmers could be excused if they were confused by these claims. When the Texas legislature offered a $50,000 reward in 1903 to anyone who could devise a method to control the insect, farmers surely recognized that the demonstration agents didn't have the answer. The whole legislative prize was a ruse from the outset. Only three weeks after it was offered, the chairman of the governor's Boll Weevil Commission, a man supposedly charged with judging the hundreds of boll weevil "solutions" sent in by Texas farmers, admitted that he had no intention of finding a winner. The commission would conclude that none of the proposed solutions was feasible and that only the cultural methods offered by the state and federal agriculture department worked to control the bug. The prize was meant to garner public attention, and then to direct that attention to the state's own solutions.[14]

Facing this bleak portrait of the weevil's threat, tenants were left with three possible solutions: quit farming, move to another farm, or continue to live and farm alongside the insect. The easiest decision a tenant could make was to simply to get out of the weevil's way. But this could mean any number of options. Not all chose to pack up everything and move out of the weevil's path. The case of one Louisiana tenant farmer sheds light on how complex the boll weevil's affect on tenants' lives could be.

At the turn of the century, B. B. Sochon was a tenant barely getting by on land owned by Louis Stelly. Sochon and his family were, like thousands of other tenants in the state, in debt; as a result he agreed to stay with Stelly for another year to try to square his balance. The local demonstration agent, L. E. Perrin, knew that both Sochon and Stelly were struggling to grow a profitable amount of cotton, but Stelly had refused Perrin's aid on a number of occasions, and the agent knew that he could not directly help Sochon, the tenant, without Stelly's permission. The great majority of landowners refused to allow agents to discuss farming techniques directly with their tenants, and Stelly was no exception. Even as the first few weevils appeared near their land in late 1905, farmers like Stelly and Sochon continued to devote most of their acreage to cotton. "They thought that they could chase weevils and pick squares on their ordinary acreage," Perrin remembered. "They all had to take their medicine at the first attempt." Over time, the boll weevil made converts out of some farmers, including Stelly.[15]

In 1907, boll weevils battered Stelly's crop, and the farmer barely broke even. The following year he agreed to work with Perrin as a demonstrator. Stelly advised Perrin, and by extension Sochon, on how much of his land to devote to cotton and how much to leave to corn and other crops. He directed Perrin to use a certain kind of cottonseed and told the farmer when to apply fertilizer. Perrin, in turn, passed this information on to his tenant. Throughout the season, the farmers worked according to Perrin's instructions. By fall they had made a bumper cotton crop. Sochon made twenty-three bales of cotton on land where he had previously made only eleven. He sold the cotton for enough money to pay his debt to Stelly and still pocket a profit of $1100.[16]

To Perrin's surprise, the day Sochon settled his debt he quit cotton farming altogether. His profitable year broke him from cotton's grip and allowed him to walk away. Not only had Sochon decided to leave Stelly's farm, he and his wife abandoned farming altogether. Sochon "decided to go merchandising," Perrin reported, "he was through eating boll weevil pie." Like Sochon, many tenants had had enough of farming under such precarious circumstances. As his experience illuminates, some tenant farmers found their way out of poverty only through the landholder's decision to seek a demonstration agent's help. Had Stelly not chosen to work with Perrin, Sochon's end-of-year receipt would certainly have been less, if positive at all, and he would have remained in debt and in place.[17]

Tenants like Sochon and McQuinney were always left out of these decisions about how to farm in the presence of the weevil, yet they performed most of the labor themselves. This frustrated most tenants, who recognized that their livelihood at least in part depended on their landholders' willingness to seek help from government agents. As Sochon's case demonstrates, some desired to quit farming altogether, but it is impossible to say just how many tenant farmers quit the fields and sought work in another position. Though anecdotal, stories like Sochon's are not rare. There were jobs available in rural towns, and landless farmers sought them out. Gins, warehouses, mills, and stores all needed laborers, and the promise of a steady wage, even if only for a part-time position, was attractive to tenants who were risking debt each year with a cotton crop.[18]

Beyond knowing how many farmers followed Sochon's path out of agriculture altogether, it is very difficult to make generalizations about the overall connection of the boll weevil to migration. Some tenants moved out of the state, some moved only one farm away, some left then returned. Families split up to move. Other groups of people combined. And of course many people had a multitude of reasons to move on, not only the arrival and influence of the boll weevil. All of these factors made tracking the movement impossible, particularly during the weevil's first ten years in the South, when it was present only in Texas and parts of Louisiana.[19]

Scholars have weighed in on this issue to a surprising degree. While economic historian Robert Higgs claimed that the boll weevil had little effect on migration patterns, Arvarh Strickland drew the opposite conclusion, arguing that the pest was an "emancipator" of southern farm labor.[20] The best effort to understand the impact of the boll weevil on migration was made by Fabian Lange, Alan Olmstead, and Paul Rhode, who found by looking at very local data that the weevil caused great disruptions in local migration patterns ahead of the boll weevil's own movement.[21] The problem that confronts all of this work lies in accurately tying a tenant's motivation for moving to that movement itself. Strickland used broad labor department reports to argue that the weevil was a factor in African American migration from the South (he did not focus his discussion on any one place or time period). Higgs, however, charted "somewhat vague" data on crop loss due to boll weevils against "fairly reliable" estimates of black outmigration from the cotton states to conclude that the boll weevil had no effect on the Great Migration, except in the early 1920s in Georgia and South Carolina.[22] Though Lange, Olmstead, and Rhode paid much more attention to migration ahead of the weevil, they too failed to take into account how tenants' decisions to migrate because of the boll weevil varied depending on time and place.[23]

Understanding the specifics of this migration—who moved where when and why—is perhaps less valuable than historians have hoped. We know that there was migration, and that tenants threatened to leave, and that they used this as a weapon to strike better deals with landowners. The extent to which owners went to guarantee their supply of labor reveals just how willing tenants were to pack up and leave. The reality that tenants who were clear of debts were free to move each year frightened landholders, who knew they could not possibly make a profit without a sufficient labor force. This pressured landowners to find ways to guarantee a labor force sufficient to work their cotton. Many turned to chicanery—cooking the books so that workers were always in debt at the end of the season—or more often simply charging exorbitant interest rates on items that croppers had purchased during the year. Tenants without family or social ties to a particular place, however, often left in spite of these debts. As a result, in places where labor was in demand, tenants' own willingness to work, and to take that labor elsewhere, constituted considerable power over landowners.[24]

In addition to outright physical violence and economic coercion, many landholders attempted to hold workers by catering to their perceived social needs. This was a carrot to coercion's stick. Some planters in Texas and Louisiana sought to make social environments where sharecroppers would want to live and work. Tenants knew that certain plantations had a dynamic social character. Recognizing that some laborers sought work on places with vibrant social lives, owners allowed, and in some cases aided, the creation of workers' recreational space in the hope that it might tie down workers.

Landowners for their part encouraged a certain amount of partying and lawlessness on their land. One white Texas planter told sociologist Paul Taylor that on his farm "there are some strays—Negro and Mexican—who gamble the Mexicans out of their earnings. There are some prostitutes. Some say [the laborers] stay better if you allow [prostitutes]." These interlopers also tended to keep workers poor, which their bosses knew would force them to work in order to earn back their wages. "Gamblers and prostitutes come and get places and pretend to pick," a landowner reported. "It is better for the farm in one sense, because [tenants] work better when they have no money." A different planter explained that the "the way to keep Negro labor is to let them have women and shoot craps." "They would rather gamble than anything, have home brew, and dance," Taylor was told.[25]

Planters also found that they could manipulate workers' tenure by managing their access to this cultural recreation. "The love of the Mexicans for dancing was frequently indulged by farmers," Taylor states, "who often provided lumber for construction of dance platforms." One large grower told Taylor, "If the Mexicans get restless, and want to leave before the cotton is picked, I tell them we are going to have a *baile* in two weeks, and give them colored soda, etc." Despite the conclusion of some historians that the parties, dances and other cultural expression by workers was "a space that whites barely touched," landowners clearly had the means to open, limit, or close that space.[26]

While planters and other white observers saw these parties as a wild atmosphere of sex, cards, and drinking, tenants understood them as more constructive, even political, social spaces. Workers let loose for sure, but on one level the impetus for and outcomes of these occasions was a grappling for power. Workers and owners *together* created this space, but it was the tenants who in the end drew strength from it. Out of this environment of celebration and recreation came a unique music fostered by and reflective of tenants' relationship to the land and their cotton-dependent livelihood. It should come as no surprise, then, that in it one hears tenants' concerns about the boll weevil in particular. These songs became powerful statements of the boll weevil myth, proclamations of what the pest would mean for tenants' lives.

Landowners may have believed they constricted tenant movement by throwing a party, but the content of the songs that tenants performed and listened to railed against the structure of control that landowners established. From stages and porches, or in the fields themselves, women and men sang of their familial, spiritual, and work lives. From a distance, landowners peeked in on these performances thinking that the dances and singing would drain the workers of money and desires to flee, rather than give them feelings of strength and community.

Songs that tenants sang for one another were loaded with their views on their place in the cotton economy and southern society writ large. In the case of these songs, the "hidden transcripts of resistance" that historians associate with the cultural production of workers turn out not to have been well hidden at all. For example, "Corrido de Texas" (Ballad of Texas), heard and transcribed by Taylor in the early twentieth century, recorded tenants' dependence on moving and understanding of the larger processes at work in their lives. In Spanish the balladeer sang:

> Goodbye, state of Texas,
> with all your growing crops;
> I am leaving your fields
> so I won't have to pick cotton.
>
> These trains of the T & P [Texas and Pacific Railroad]
> that cross Louisiana
> carry the Mexican
> to the state of Indiana
>
> Goodbye, Fort Worth and Dallas,
> cities without a lake;
> we'll see each other when I return
> from Indiana and Chicago.[27]

Songs like this reflect not only workers' attachment to the cotton crop and their understanding of the nationwide system of labor in which they operated; they also reflect tenants' appreciation of the power of their own movement. The "Corrido de Texas" is at its heart an assertion of workers' control over their own fate. It reflects discontent about the physical work of cotton picking and the knowledge that there were better jobs elsewhere. (The final verse also demonstrates the reality that migration was seldom final for tenants; the singer admits he or she will make it as far as Chicago, but will return.)

When folklorist Gates Thomas traveled south Texas to record the songs of tenant farmers, it didn't surprise him to hear workers singing about their conditions, but it must have struck him as odd to find a song about the boll weevil only five years after the bug's identification. In the 1890s, Thomas, an English professor at Southwest Texas State Teachers' College, made trips around the state to study what he called "Negro work songs." He approached black tenant farmers and asked them what songs they sang about their work in the region's cotton fields. On an 1897 trip he heard one about a bug. The singer began:

> The boll-weevil says to the sharp-shooter, "Pardner, let us go,
> And when we strike that cotton patch, we'll take it row by row;
> For it's our home, Babe, for it's our home."[28]

(The sharpshooter was another cotton insect pest.)

To Thomas's white ears and sensibility, "The Boll-Weevil" was merely a humorous folk tale, "imaginatively true to the time and region in which it arose." The professor eventually published his analysis of the song in the 1920s, after gathering several additional verses, but he never took it seriously as a record of tenant life. The song's singers whom he had interviewed were "'lusty, phallic, Adamic' Negroes of South Texas, shiftless and shifting day laborers and small croppers who follow Lady Luck, Aphrodite, and John Barleycorn." To Thomas, these workers—he called them "kinkies"—moved around the cotton South not in reaction to the realities of tenancy, familial bonds, or agricultural change, but to follow luck, sex, and booze. For Thomas, it was these tenant farmers' ignorance of farming that made their work songs "the most authentic."[29]

It is not surprising, considering the folklorist's view of black Texans, that he paid little mind to the noninsect characters in the boll weevil song, and ignored the possibility that the singers might be using this tale of cotton de-struction to say something larger about the agricultural society in which they worked and lived. He ignores the song's representation of labor and the dual-ity of meaning in its themes of defiance and mobility. Like many of the white intellectuals who studied the song in the late nineteenth and early twentieth centuries, Thomas missed the point.[30]

The version of the boll weevil song Thomas encountered in the 1890s moved as sharecroppers did. The folklorist had discovered a cycle of song creation at work. Musicians made up stories about the pest and set them to music, playing them publicly for cotton workers. Field laborers and musicians—groups that were often one and the same—then moved, taking the songs about the boll weevil with them to a new place, where the real-ity of the pest's effects were different. The songs were reborn, replayed, and eventually reached new audiences and the cycle repeated. Each boll weevil song became part not only of cotton laborer's memories as they moved, but of professional musicians' repertoire as well. As a result, audiences far from weevil-infested cotton heard the musical news of the pest. It's a point that bears repeating. Men and women, particularly rural black southerners, heard of the boll weevil's history through these songs long before the insect had even crossed the Mississippi River.[31]

The spread of the boll weevil song from Texas was due not only to a migrant cotton tenant workforce, however, but also to a vibrant commercial musical community as well. Several of the twentieth century's most important American musicians were raised in Texas and Louisiana at the moment that the boll weevil pushed through. The "Father of Ragtime," Scott Joplin, was born on a cotton farm in northeast Texas around 1868. His family, like so many other migrant laborers, moved to Texarkana once the railroad had been built there, and there Joplin developed an interest in music. As he grew, Joplin cut his musical teeth in the fraternal halls and social clubs of Texarkana, where he probably learned and played songs about the boll weevil. Rural African Americans who rode into Texarkana to buy necessities and socialize often attended the dances where Joplin played. There was a great deal of cultural exchange as city and rural people talked, played, and danced, and Joplin undoubtedly encountered boll weevil stories and perhaps heard a song or two about the pest. By his twentieth birthday Joplin himself packed up his arsenal of songs and developing talent and hit the road, spreading the songs he knew first to the West, then up the Mississippi River to St. Louis.[32]

Not far from Joplin's birthplace, Blind Lemon Jefferson was born in 1893. Blind at birth, Jefferson began playing guitar as a young boy, but by his early twenties he was making a living as a musician on the streets of Dallas. Around the same time Huddie Ledbetter, later known worldwide as Leadbelly, was born in the borderland region between Louisiana and Texas. By the time he was old enough to play guitar, he was rambling around his hometown playing on the street, at house parties, and in clubs listening to the stories and songs of workers escaping from work in weevil-plagued fields. By the time he left Moorinsport, Louisiana, in 1906, he had learned "Boll Weevil" from his Uncle Terrell, who lived in west Texas. Leadbelly took that song and dozens of others with him as he traveled to Shreveport and later Dallas. There, he hooked up with Jefferson and a third bluesman, Josh White. The three taught each other songs they had picked up and in turn played all over town. They eventually became an important blues triumvirate, traveling independently on from Dallas all over the world, spreading not just their unique sound but also the ever-changing story of the boll weevil.[33]

So what were the songs that these musicians heard, sang, and spread throughout the South? Was the boll weevil a simplistic figure in a larger story or did the pest come to represent something deeper to these rural people? Many who heard and sang these boll weevil songs found in the cotton pest a kindred spirit. The songs' narrators viewed the pest's invasion and slow, unpredictable spread as a frenetic movement not unlike tenants' own. Themes of migration and longing for better work are prevalent in the collection of boll weevil songs that sprang up in the infested territory almost as soon as the pest began destroying south Texas cotton.

In the version Gates Thomas first heard in the 1890s shortly after the weevil's identification in south Texas, the singer opens with an observation reminiscent of Charles DeRyee's 1894 letter to the USDA. The speaker recalls:

> The first time I seen him he wuz settin' on a square;
> Well, the next time I seen him he wuz a-crawlin' everywhere,
> Just a-huntin' him a home, Babe, just a-huntin' a home.[34]

The pest's ability to multiply, seemingly right before the farmer's eyes, was impressive, but its motive for reproducing and wandering ("he wuz a-crawlin' everywhere") suggests the more important point. The boll weevil was on the move not simply to destroy cotton but to find a home. Common to almost every version of the boll weevil song recorded (either in text or audio) from the 1890s to the present is the boll weevil's preoccupation with the home. In almost every example, the insect is finding a home in a cotton field, invading someone's home, forcing someone from a home, or most commonly "just lookin' for a home."[35]

Some versions of the song offer a detailed explanation of where the insect had been and where it was headed:

> The boll weevil is a little black bug
> Came from Mexico they say,
> All the way to Texas
> Just a-lookin' for a place to stay
> Just a-lookin' for a home, just a-lookin' for a home.[36]

Whether or not the songs explained where the weevil had been, most name precisely what home it was in search of: "yours." "Have you heard the lates', the lates' all yo' own?" a black sharecropper sang for Thomas in 1906. "It's all about them weevils gonna make yo' fa'am [farm] their home." In these versions, the weevil was more than just a fleeting danger; it was a threat to the very livelihood of the tenant farmer.[37]

As Thomas traveled Texas in later years he heard new versions of the song, reflecting farmers' varying strategies to fight the beetle. One rendition referred to the pest's ability to beat any farmer's strategy to kill it. Poison, weather, heat—nothing could stop the weevil and its search for a home:

> So they took the little boll-weevil and put him on the ice.
> He sez to the farmers, "I say, but ain't this nice!
> But it ain't my home, though; no, it ain't my home."

> Then he took the little boll-weevil and put him in hot sand.
> He sez to the farmers, "Will, and I'll stand it like a man,
> Though it ain't my home, Baby; no, it ain't my home."[38]

This apocryphal scene, which Thomas heard sung by black sharecroppers in east Texas, shares the imagery and narrative of the tale Seaman Knapp heard an old farmer tell when he first visited the same weevil-plagued area in 1903 (the scene that opens chapter 1). That farmer had tried poisoning, drowning, and burning the pest, which refused to perish. In the song, the boll weevil is again all-powerful and this time can speak, telling the farmer that indeed he can survive the burning sand "like a man."[39]

Leadbelly's version of the "The Boll Weevil," recorded when the singer was on death row in Louisiana in the early 1930s but almost certainly picked up during his youth in Texas, also painted the pest as a kind of trickster. After singing verses about the pest's ability to defy ice and heat, the narrator admits that, in the end, it is he—the sharecropper—and not the landowner who will suffer. Once the insect has beaten back the narrator's attempts to destroy it, the tenant farmer is left with a single bale of cotton, which he owes the merchant to settle his debts:

> Now the farmer, he said to the merchant,
> "I never made but one bale, before I'll let you have that last one,
> I will suffer and die in jail,
> I will have a home, I will have a home."

The cropper has faced certain debt for another year, and realizing that, instead of settling with the merchant for a paltry amount, he refuses to pay off his debt, which will land him in lockup. At least in jail the sharecropper will, like the boll weevil, have a home.[40] Like the tenants in Texas and Louisiana, Leadbelly's speaker had limited options for dealing with the pest's destruction. In the final verse Leadbelly's tenant farmer settles on the most realistic choice, to pack up and move on. Like the boll weevil, ever moving in search of more cotton, laborers migrated from one landlord arrangement to another, in search of a place with economic opportunity and social independence.[41]

As revealed by Thomas's textual recordings of boll weevil songs made in the 1890s, as well as the published tunes that came later from singers who had cut their musical teeth in weevil-plagued Texas and Louisiana in the early twentieth century, the tunes' origins lie with the men and women who first encountered the pest on cotton's frontier. These boll weevil blues reflect not just life with the bug, but tenants' place in cotton culture itself. The versions recorded in Texas and sung by Texans are some of the most straightforward takes on the song. The stanzas proceed in a linear and chronological fashion. As the song moved, of course, it changed. New singers changed the words and music to reflect local realities. As we'll soon see, the first versions recorded in the Yazoo-Mississippi Delta are much more complex, even mysterious, reflecting the weevil's new threat to that cotton-dense region.

*

By the end of 1907, boll weevils could be found in five states (Texas, Oklahoma, Louisiana, Arkansas, and Mississippi), but its appearance had produced varying levels of damage. (See Figures 7 and 8.) In total, the pest had destroyed more than three million bales worth $200 million. In its most damaging year to that point, the weevil had destroyed 7.7 percent of the South's entire cotton crop, but there had been much more destructive years in certain locales. In 1903, the weevil reduced Texas's crop by 19.4 percent, and in 1907, it took 20 percent of Louisiana's crop. By 1908, Arkansas, Oklahoma, and Mississippi had experienced only very small crop losses.[42]

Despite the insect's limited location and the local variations in its ability to destroy the precious fiber, by the turn of the century a range of people had already spread news across the nation of the insect's voracious appetite and its wholesale transformation of cotton farming in Texas and Louisiana. The reality of the weevil's effects was more complicated, however. From 1890 to 1910, Texans had more than doubled their cotton acreage from under 4 million to nearly 10 million acres. Their production too increased from 1.5 million bales in 1890 to 2.5 million bales in 1910. These increases are due in small part to farmers following the USDA's cultural recommendations, but a much more important factor was the westward expansion of cotton farming itself. Thousands of farmers looked to the western, drier areas of Texas, where the

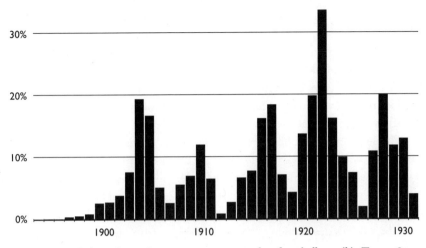

FIGURE 7. Estimated annual percentage cotton crop loss from boll weevil in Texas, 1892–1930. Illustration by the author. Data from Willard A. Dickerson et al., eds., *Boll Weevil Eradication in the United States through 1999*, the Cotton Foundation Reference Book Series, no. 6 (Memphis: Cotton Foundation Publisher, 2001), 614–15.

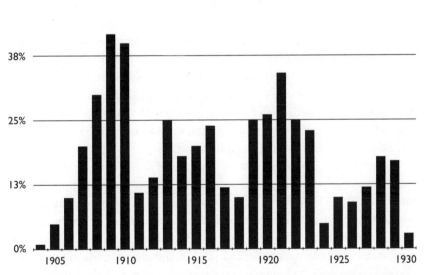

FIGURE 8. Estimated annual percentage cotton crop loss from boll weevil in Louisiana, 1904–1930. Illustration by the author. Data from Willard A. Dickerson et al., eds., *Boll Weevil Eradication in the United States through 1999*, the Cotton Foundation Reference Book Series, no. 6 (Memphis: Cotton Foundation Publisher, 2001), 614–15.

weevil was less of a factor. In Louisiana, however, there was no vast western prairie on which to extend cotton cultivation when the weevil arrived there in 1903. The pest's effect on cotton acreage was much more pronounced as a result. From 1900 to 1910 Louisiana farmers planted 30.5 percent less land in the staple crop. Production fell even more. Over the same period Louisianans harvested nearly a half million fewer bales of cotton, a 62 percent drop. As Figure 8 shows, much of this was due to the boll weevil, which had more fruitful years as the decade progressed.[43]

Though clearly the boll weevil had affected local agriculture in Louisiana, the reality was a long way from the myth. By 1910, farmers, national and regional newspapers, USDA officials, local demonstration agents, politicians, and travelling singers—both people who had experienced boll weevil conditions and those who had never been near the beetle—were loudly arguing that the weevil's destruction of cotton had destroyed all farming in Texas and promised to do the same in every state that it encountered. Knapp himself wrote of "whole towns deserted" in 1903, and his agents, who were now fanned out into five states, capitalized on this dire picture.[44]

These agents, still numbering only in the twenties in 1903, were spread increasingly thin across the boll weevil territories. (The original legislation that had funded their work limited their travel to places where the weevil was present.) As the pest moved, their territory grew. In 1905, the General Education Board, John D. Rockefeller's multimillion-dollar organization founded to aid southern education, began giving significant money to Knapp's Farmers' Cooperative Demonstration Work for the purpose of hiring more agents. These new agents were sent ahead of the weevil front to prepare farmers for the coming pest, as well as to teach generally applicable modern farming methods. By the time the boll weevil first reached Mississippi in 1907, Knapp still had fewer than fifty agents, but in 1912, thanks to the support of the General Education Board, he had seven hundred.[45]

While the demonstration system had spread throughout the boll weevil territory and to points ahead, it was hardly the only organization fighting the insect. Rural reformers had allied with scientists, farm educators, and businesses dependent on cotton, such as railroads and seed companies, and were themselves travelling the countryside talking to farmers, publishing pamphlets, and holding meetings in schools and churchyards to preach the USDA's cultural method of boll weevil control.

This strategy—defined by early season planting of fast-ripening seed, using plenty of fertilizer, and picking the crop as early as possible—hadn't changed much since 1895. As the pest moved into new areas of the South, however, scientists began to understand how different local environmental forces aided or stifled the pest's ability to damage cotton. The weevil did more damage in rainy years, they found, and on land where there were ample hibernation spots. To observers in the Mississippi Delta, the flat, wet, warm, cotton-packed land just ahead of the weevil, this new information brought only disquiet.

CHAPTER 3

"Map Maker, Troublemaker, History Maker": The Boll Weevil Threatens the Delta

When the Southern Railway's Special Agricultural Train pulled into tiny Belzoni, Mississippi, on March 10, 1909, two hundred people had gathered to welcome it. Down the track in the town of Richey, over a hundred farmers came in from the countryside to meet the train, despite the fact that the settlement's population was only thirty. Later that day the train stopped in a place so small that news reporters were not sure if it was called "Swiftwater" or simply "Swift." Confusion over its name stemmed from the fact that it had a train stop, but no depot, bank, or post office. Despite the diminutive town's limits, three hundred people greeted the train. At stops in the little villages of Arcola and Hollandale, 150 tenant farmers and small landowners showed up. The crowds in each of these towns were less interested in the train itself than its passengers, the state's top farm experts. Rural Mississippians were anxious to hear their plans for combating the slowly approaching boll weevil.[1]

Aboard the train were professors from Mississippi Agricultural and Mechanical College (Mississippi A&M), scientists representing the USDA and Mississippi's Department of Agriculture (MDA), and the editors of several farm newspapers. These men rode through the Mississippi countryside on the eve of the boll weevil's entry into the Delta doling out advice. They talked to farmers about crop diversification, hog raising, applying fertilizer to corn crops, and the benefits of home canning, but the subject that brought thousands of people in from their farms to crowd onto train platforms was the impending arrival of the boll weevil. A few weeks earlier the Illinois Central had organized a similar excursion, dubbed the Boll Weevil Special, through the

Delta, loaded with many of the same experts to warn farmers of the potential damage of the insect invasion.[2]

Locals' attention to the Southern Railway's train impressed both the experts on board and those journalists sent to cover the tour. After the coterie's first day of stops, the Memphis *Commercial Appeal* reported "several hundred farmers" had been "anxious to learn" from the "experts." The New Orleans *Daily-Picayune* described "planters in the great Delta counties . . . eager to learn of advanced farming." The experts themselves deemed the first two legs of the three-day tour wildly successful. The size of the audiences, even in the region's smallest towns, had exceeded their expectations.[3]

On the final day of the train's tour, however, things changed. On the morning of March 11, it made the first stop of the day in Greenwood, the second largest city in the Delta. Greenwood was the Leflore County seat and home to nearly six thousand people. When the train pulled into town, however, the professors and agriculturalists on board were shocked to find the station empty. As the passengers disembarked in search of the throngs that had met their previous stops, a small body of men approached. After a brief meeting with this group—which a witness would later refer to only as "business men" and "Greenwood people"—the professors, scientists, and editors hurried back into the cars, and the train fired up and moved down the track out of town.[4]

Where were the crowds of farmers anxious to find out about the boll weevil? What explains the Greenwood group's unwillingness to allow the train's experts to speak? The answers to these questions lie in the complex relationship between the town's business interests, state agriculture experts and scientists, planters, and labor. The catalyst, however, was the boll weevil itself. Still physically miles from the Delta's cotton lands, the spread of stories about the insect's voracious appetite and destruction of plantation agriculture had put the pest in the forefront of the minds of those in power there.

The group of men who met the train had gone to great lengths to ensure that not only would no crowd be assembled to meet it, but that not a word about the boll weevil would be uttered. The group had suppressed publicity of the train's program. They had torn down all advertising posted by the Southern Railway in Greenwood. They had refused to let the experts speak publicly about the weevil. Greenwood's planters and merchants conspired to put a lid on any discussion of the approaching pest.[5]

A railroad agent later recalled that despite the huge crowds that met the train in the region's rural sections, "about Greenwood and Greenville," the anti-education sentiment "seems to be crystallized." Explaining the events on the rail platform that morning, the agent wrote that planters "view with much concern any discussion or agitation of the boll weevil menace." "The Greenwood people had reached the conclusion that our train was *featuring*" the boll weevil, and as a result, the group "suppressed our advertising and kept all reference to the meeting out of the local papers." The Delta elite "fear[ed] dire consequences" if these professors and crop experts were allowed to speak.[6]

The incident raises the question: what consequences were more dire than the boll weevil? After all, these educators were bringing solutions for the pest problem, so why would these Delta planters and businesspeople want to suppress this knowledge unless they feared something even greater than their cotton's destruction by the weevil?

As the story of collusion and suppression surrounding the Greenwood train stop demonstrates, in the face of the boll weevil threat, the future of the Delta was up for grabs. The myth of the weevil's transformative force on southern cotton had reached the Delta long before the insect itself did. By the turn of the century, both planters and laborers not only knew the insect was coming, they knew a good deal about it—enough to be scared that the pest would ruin this cotton plantation kingdom. As the beetle arrived in late 1907, each group sought to use the fear of the weevil to their own advantage.

The weevil forced planters to tighten their grip on Delta society, meaning everything from ownership and control of farmland to the movement of people, credit, and even knowledge. They also sought to control the physical growth of state and federal agriculture programs—where the research was done, how it was performed, and who did it—as well as the actual information about the boll weevil that these researchers and educators disseminated. Second, planters tried to control the boll weevil threat by controlling labor. By attempting to limit what tenant farmers knew about the insect, landowners hoped to avoid a mass exodus of workers. Ironically, as planters were using the boll weevil as an excuse to further tighten the screws on tenant farmers by limiting what they knew about the pest, they were also using the weevil's arrival as an excuse to examine alternatives to the very labor system over which they obsessed.

The region's tenants, however, recognized that the boll weevil was a threat to this planter hegemony and were, to a limited degree, encouraged by the instability the pest promised to bring. Many looked to use the insect's presence as a chance to improve their social and economic livelihoods by seizing access to lands made less valuable by the weevil, by retooling their contracts with landowners, or by moving in search of a better deal. Extension agents too used the idea of the beetle's destructiveness to get a foothold among the Delta elite, to whom they sought to bring modern, diversified farm techniques.

The most important motivating force for all three groups was not the weevil itself, but the *notion* that it was unstoppable. The irony is that the boll weevil had little long-term effect at all on the ability of Delta farmers to make healthy, profitable crops of cotton year in and year out. From 1900 to 1930 farmers there more than doubled their cotton acreage and bale production.[7] There was little demographic change, either. From the moment of the boll weevil's arrival in 1908 to the start of the New Deal, the same white landowners exercised their social, political, and economic power over black sharecroppers by using their control of the Delta's physical environment as their main weapon.

*

The Yazoo-Mississippi Delta is a swath of land that begins at the Mississippi-Tennessee border on the north, and runs south 225 miles to Vicksburg, Mississippi. More than 5.5 million acres make up this Delta, which is confined by the Yazoo River on the east, and the Mississippi River on the west. At the area's widest point the rivers are sixty miles apart. Though the region is defined geographically by the rivers that lend it its name, the Yazoo-Mississippi Delta is not a proper river delta at all, rather the confluence of the two rivers. (Throughout this chapter and the next, I use "the Delta" to refer only to the Yazoo-Mississippi Delta, which lies entirely in the state of Mississippi.) It was the region's connection to these waterways that initially gave the Delta its link to the world and hastened its late-nineteenth century rise as a cotton kingdom.[8]

From the mid-1800s to 1900, black labor and white capital transformed the Delta forever. Fewer than fifty years prior to the boll weevil's arrival, there were only a scattering of settlements in the region and a small population; the bulk of the land was uninhabitable swamp. As Mikko Saikku argues, even as late as the end of the Civil War, "the prospects for economic development were meager."[9] In 1860, only 10 percent of the Delta was cleared. Civil War soldiers found upon their return in 1865 that a few of the small Delta towns built prior to the war had simply disappeared. Clarksdale, which would become one the region's most populous towns, did not even have a formal street layout until 1868. Though mostly uninhabited by humans, the land was not vacant. In addition to the panthers, bears, and other animals that roamed the marshes, huge bottomland hardwood trees still covered most potential cotton fields. The landscape would require tremendous labor if it were to become a cotton kingdom.[10]

In the late 1800s, despite the impediments of flooding, dense flora, disease, and remoteness, men began clearing and draining the land to prepare it for cotton. Adventurous capitalists, armed with thousands of dollars in northern capital from banks and railroads, bought up huge tracts of Delta land, gambling that these lowlands could be freed of trees and standing water and made tillable. It was a significant risk, but the payoff, they believed, would be enormous because of what lay under these pioneers' feet. The rivers that formed the edges of the Delta's boundaries had flooded thousands of times over millennia, leaving behind a nutrient-rich soil unlike any other in North America. Topsoil, that crucial layer of bacteria and minerals that fuels any vegetation for its growth, was (and remains) key to all agricultural pursuits. Historian Steven Stoll has written that "soil is a bank account for fertility that farmers draw upon, and the balance is always low." Not in the Delta. Flood after flood had made the region's topsoil unbelievably deep, and by extension the land was amazingly fertile. "The river left gold in the Delta," wrote John Barry. "Elsewhere one measures the thickness of good topsoil in inches. Here good lush soil measures tens of feet thick." This soil, along with a long, warm growing season, and flat, even land would make the Delta a nearly ideal place to build a cotton kingdom, if only the trees and water could be permanently removed.[11]

By the end of the nineteenth century, the Delta land grab was on. Will Dockery's story is typical. In the 1880s, with a thousand-dollar gift from his father, Dockery left his family in Memphis and moved to the Delta to make himself into a cotton planter. He bought land east of Cleveland and enticed labor to help him clear the trees and brush. Removing the stumps and cane proved too tough for many of his neighbors, and Dockery soon found he could purchase nearby land in exchange for livestock and guns, rather than cash.[12]

As his holdings grew, Dockery found life in the Delta was no pastoral wonderland. Dockery's son, Joe Rice Dockery, recalled growing up in Memphis and traveling south to see his father in the Delta only for Christmas, vacations, or to "hunt frogs." Will, meanwhile, acquired and cleared more land. By the mid-1890s, the Dockery plantation employed some clerks, storekeepers, a few wage laborers, and hundreds of sharecroppers. He had built a house, cotton gin, store, and dozens of cabins and outbuildings. By the time the boll weevil arrived in 1909, the Dockery conglomerate even printed its own money to pay its workers.[13]

Capitalist adventurers like Dockery attempted to create in the late-nineteenth-century Delta a modern world with up-to-date business and accounting practices, directly connected to the global cotton economy, but fashioned with the appearance of a mythical antebellum tradition. James C. Cobb describes the spirit of men like Dockery:

> Fancying themselves heirs to an aristocratic antebellum tradition, this cadre of white leaders sought to create through an ironic combination of economic modernization and racial subjugation a prosperous and politically insulated cotton kingdom where the Delta planter's longstanding obsession with unfettered wealth and power could be transformed from Old South fantasy to New South reality.

Despite their intentions, these white leaders had various degrees of success at mixing the Old South ideal with the modern, industrializing New South. What Delta landowners were unwaveringly good at, however, was growing cotton, and they did so by not only making the plant grow, but by engineering an entire society—a physical, economic, and social environment—that revolved around the staple.[14]

The key economic link that allowed the Delta's environmental transformation to work economically was the arrival of the railroads. Though a rudimentary rail system first allowed Deltans to ship cotton to regional markets in the 1880s, the system was inefficient. Planters like Dockery, John M. Parker, and LeRoy Percy helped subsidize the development of branch lines and were rewarded in turn with stops at each of their plantations and gins. In 1892, the powerful Illinois Central railroad bought out most smaller rail lines in the Delta and renamed its system the Yazoo and Mississippi Valley Railroad (YMVR). The YMVR soon became the most profitable segment of the Illinois Central's vast rail network. By the turn of the century, ginned cotton could be loaded onto a train at a remote plantation and carried by rail to New Orleans or Memphis and sent from there to the mills of the northern states or England.[15]

It was no accident that the evolution of the railroads corresponded to the rise of the planter class. The biggest planters served on the boards of railroads and banks. And railroads themselves were not solely in the transportation business, either; these companies owned entire plantations and were knee-deep in federal, state, and local politics. As a result, the lines between categories of interested parties — planters, shippers, bankers, industrialists — blurred as cotton became increasingly profitable. At root, the reason for this economic incest was cotton. The combination of the soil's nutrition, the Delta's long hot summers, and the efficiency of plantation operations had made the region the center of U.S. cotton production, and it had made the fiber the center of the region's culture.[16]

Cotton was more than an industry, wrote one observer of the Delta for the *New York Times*; "it is a dynastic system, with a set of laws and standards always under assault and peculiarly resistant to change. It is map maker, troublemaker, history maker." The region's residents committed to cotton to the exclusion of all else. There was no promising industry in the Delta that fell outside the purview of its production. "Cotton is more than a crop," wrote Delta writer David Cohn,

> it is a form of mysticism. It is a religion and a way of life. Cotton is omnipresent here as a god is omnipresent. It is as omnipotent as a god is omnipotent, giving life and taking life away. Here the industrial revolution is an academic adumbration dimly heard, an alien device scarcely comprehended.

Despite Cohn's catchy prose, the industrial revolution had made waves in the Delta; the cotton plantations themselves were industrial marvels operating with modern methods of accounting, organization, and labor control. In fact, the supremacy of the fleecy fiber in Delta culture was a reflection of just how industrial the process of cotton production had become. The successful implementation of these modern methods on plantations relied on the regional economy and society's complete acquiescence to the crop. By 1900, Deltans' minds had been fixed on the plant, the land had been transformed, the transportation system built, and credit extended. To the region's planters and business elite, the only unknown factor buttressing the region's agricultural preeminence that remained unknown was the behavior of the non-landowning workforce.[17]

As in much of the South, landlords and tenants in the Delta developed and revised new, flexible, and constantly changing systems of tenancy. In the Delta, however, the necessity of maintaining a workforce sufficient to clear more land and crop the fields was greater than in any other southern region. After all, planters had relied on thousands of black men and women to remake this marshy land into a cotton paradise, but they would still need labor to work the crops every year. As a result of this basic, profound need, a relatively distinct form of sharecropping emerged.

Though the sharecropping system certainly benefited planters, savvy sharecroppers realized that if their share of the crop was sufficient, the landlord was honest, and his or her credit terms not too outrageous, that it was quite possible to show a profit at the end of the season. Most laborers, as a result of this system, tried to work the most productive land under the most equitable terms. In other words, the condition of the land, its fertility, drainage, topography, and location, determined its value to sharecroppers. This pulled thousands of laborers to the Delta's rich flatness. Many were also "pushed" from the worn-out cotton lands of Georgia and the Carolinas and from the boll weevil–infested fields of Texas and Louisiana. *The Progressive Farmer* reported in 1900 that Georgia landowners faced a severe labor shortage "owing to the fact that the negroes in large numbers are leaving for Mississippi, Arkansas, and Louisiana, where there is said to be a big demand for negro labor." These men and women thought the Delta's natural gifts would allow them to clear a cotton surplus at the end of the season and perhaps raise them out of poverty. The reality of working in the Delta, however, was seldom as promising as sharecroppers hoped.[18]

In addition to the Delta's attraction to cotton laborers from other parts of the South, race also set apart the region's labor system. While white sharecroppers were common in most of the South, they were rare in the twentieth-century Delta. In 1900, the region possessed the highest concentration of black labor in the U.S. By 1913, 88 percent of rural dwellers in the Delta were black, over 95 percent of tenants were black, and nearly 95 percent of black farmers were tenants. Tenancy almost without exception meant blackness and blackness meant tenancy. The region was home to a fully articulated form of racial capitalism where the system of labor control was a system of racial control.[19]

But the racialized nature of farm labor did not mean that African Americans' conditions in the Delta were completely determined by white landowners. Tenants had weapons of resistance at their disposal, most importantly movement. The constant need for labor forced landowners, merchants, and other creditors to be innovative with the profit sharing aspect of sharecropping. The relationship between cotton and credit was a chicken-and-egg phenomenon: the more cotton planters wanted to grow, the more credit they required, forcing them to plant more and more cotton. This spiraling reliance on the fiber affected not only those at the top of the Delta's economic system, but those at the bottom as well. Planters needed more labor to cultivate more cotton. The result was the crop lien system, a cornucopia of credit arrangements between workers, landowners, banks, merchants, buyers, and factors, all geared towards getting cotton into the ground and its fruit to the gins of the South and the mills of the world.[20]

On the verge of the weevil's arrival in the Delta in 1907, white planters recognized and resented how these credit systems made them dependent on African Americans. The powerful planter and politician Walter Sillers wrote to a friend that year, "I am too busy to write long letters these days what with niggers and cotton, and 'future cotton,' and law and niggers and mules, and the Lord only knows, I am kept too busy to eat or sleep much either." His repetition testifies to his frustration. What Sillers and his neighboring planters realized was that the boll weevil was more than just a threat to Delta cotton; it was a danger to their power and to the very fabric of the social and economic relationships in the region. The boll weevil might as well have been a devourer of paper money, or credit, or tenant contracts as of the cotton plant. As the pest stalked eastward, the people of the Delta, each in their own way, tried to imagine solutions to the problem that fit within their preconceived systems of social, economic, and agricultural power.[21]

*

As early as 1903, Delta newspapers published horrifying accounts of boll weevil–plagued Texas. "A Boerne, Texas, correspondent," reported the Vicksburg (Mississippi) *Herald*, "writes instructively of the Mexican boll weevil . . . a tough elusive little insect, hiding beneath the boll shuck, secure from poisoning powders, sprays, etc." In Memphis, the Delta's de facto business and social capital, the *Commercial Appeal* gloomily reported in early 1909 that the boll weevil was advancing quickly through the cotton South, "leaving ruin and disasters in its path." The Greenwood *Commonwealth* claimed that in Louisiana, "boll weevils increased . . . and cultivation has stopped." A month later it reported, "The boll weevil has spread over the entire state [of Louisiana] and is destroying practically all of the cotton forming now." Another Delta paper echoed the concern, citing reports from Texas that "weevils are numerous in timbered sections" and from Louisiana that "complaints of the boll weevil are numerous and serious." By late 1908, the papers reported on the advancing boll weevil in nearly every issue; on October 30, 1908, for example, the Greenwood *Commonwealth* published three separate articles about the pest in a single issue.[22]

Despite these dire reports, many editors gave voice to a different outlook: blind optimism. The *Commonwealth* predicted that after the hot summer of 1909, "there is no possibility that there be many weevils left next year." The Woodville *Republican* assured its readers that harmless cocklebur weevils, not cotton boll weevils, were the pests about to enter the Delta. Other news accounts expressed a strong faith in science to kill the boll weevil before it reached the Delta. Though the Vicksburg paper admitted that "the boll weevil seems a difficult subject and a serious menance [*sic*]," it predicted the problem pest would "be circumscribed by science" before reaching Mississippi. The concern about the insect's advance was enough for the paper, however, to implore Deltans not to import any seed or cotton from the weevil-infested territory. In sum, these articles underscore the degree to which the boll weevil myth was alive and well in Mississippi long before the pest itself was. Though predictions of how the insect would transform the Delta varied, it was a topic on everyone's mind.[23]

The scattered impressions of the weevil found in the pages of local newspapers were not a great improvement on the scientific appraisals offered to farmers by experts and officials at the federal level. During his 1904 State of the Union address President Theodore Roosevelt told Congress that "in Guatemala [the weevil was] being kept in check by an ant, which has been brought to our cotton fields for observation. It is hoped that it may serve a good purpose." The optimism didn't end with the president. In the USDA's 1906 *Yearbook*, entomologist W. D. Hunter revealed that experiments with cotton in the weevil territories of Texas had produced bumper crops, though at the same time he explained that this result was more likely from dry summers than from any technique or technology. In fact, these large yields, Hunter argued, "have given the erroneous impression of the prospects" of annual large crops despite the presence of the boll weevil. Downplaying the department's own successes, Hunter referred readers to the map of the pest's spread, suggesting that the wetter alluvial areas of the Mississippi Delta might be the weevil's promised land. Delta planters who paid close attention to the USDA's work found themselves in the middle of an exchange of conflicting information.[24]

At the state level, the MDA's farmers' bulletins totally ignored the boll weevil altogether until early 1906, a year before USDA entomologist Walter Hunter found the first weevil in Mississippi, six miles south of Vicksburg.[25] With the pest now present in the state, the department's efforts to educate farmers changed radically. That year the MDA began publishing concise "Entomological Press Circulars" to broadcast the latest information about the pests' location to farmers and local newspapers. In 1907 W. L. Hutchinson, director of the state's experiment station in Starkville, authored a bulletin assuring farmers that "the boll weevil does not prevent the growing of good crops of cotton." Despite the slow but steady spread of the pest through the state, Hutchinson was upbeat. He recommended farmers choose a balanced approach to fighting the pest:

> Success depends essentially on good tillage, proper fertilization, the planting of good seed of a good early variety of cotton as soon as weather conditions are favorable; and, properly cleaning up the farm of hibernating places for the weevils during the winter months.

Hutchinson's advice amounted to basically the same cultural methods first prescribed by the USDA ten years earlier. Though he could not offer anything new, or specifically geared to Mississippi farmers, he did warn readers to be prepared. "Let [the boll weevil] find the farmer ready for him," he wrote, "and his first injuries will neither be so great nor so easy."[26]

It is doubtful that any planters' fears of the weevil were assuaged by these conflicting reports coming from the assorted state and federal agencies charged with fighting the bug and educating growers, and it would have mattered little anyway. If the Delta elite believed in one thing it was this: In the half-century before the weevil's arrival, they had created, with the help of northern capital and black labor, a unique agricultural empire geared to the industrial production of cotton. This experience with the Delta's landscape had by the turn of the century nurtured a distinct and powerful ideology in the region's landowners. Planters admitted that even more important to this physical transformation than their capital and collective managerial skill had been their reliance on, and ability to control, a large black labor force. Managing these workers had not merely been one aspect of the creation of the modern Delta; to planters it had been the fundamental one. Dealing with workers had always been more important to their fight against the natural world than their understanding of forests, soil, finance, or even the cotton plant itself. Under threat by the advancing insect pest, then, these elite landowners thought first of the farm workforce.

*

Alfred Holt Stone saw his own plantation as a kind of monument to both man's power over nature and to white supremacy. Stone operated a large farm near Leland, practiced law in Greenville, and was an accomplished amateur historian and economist. Stone believed not only in the uniqueness of the Delta environment, but also in the practical application of scientific research. He turned to science around the world as a means to understand not only farming, but race as well. It is no exaggeration to say that Stone was obsessed with the differences between black and white. (His collection of printed materials on race, which he donated to the Mississippi Department of Archives and History, numbers more than three thousand items in 112 volumes.) His expertise in "the negro problem" came not only from his vast collection of "scientific" publications, but also from his "lifetime spent in the 'blackest' of the south's 'black belts.'" (Stone's conflation of soil type and skin color reveals how "natural" he viewed white supremacy.) His life as a planter had afforded him "the most constant and intimate association" both with his black neighbors and with the Delta environment. Stone's own answer to the boll weevil problem would combine these already closely linked visions of nature, science, and race.[27]

For Stone, the black labor force and the physical environment of the Delta were equally important to its future as a cotton kingdom. He knew them to be inextricably linked, almost one and the same. He rarely spoke of African Americans without mentioning nature, and vice-versa. In a 1902 paper published by the American Economic Association, Stone explained how the natural gifts of the Delta's physical world and the "prosperity" of the region's black workforce were mutually supporting factors. Just as the physical landscape of the Delta "differs radically from the rest of Mississippi" due to the "layers of fatness" in its soil, Stone wrote, it is also different "wherein the negro is immediately concerned." Stone admitted that the Delta owed its environmental transformation to the work of black Mississippians. "Its forests have been cut out by the negro," and the levees "erected mainly by the negro." While "[t]he capital, the devising brain, the directing will, constitute the white man's part," Stone conceded, "the work itself is the negro's." Relying so heavily on black labor, however, did not come without its problems, but for Stone these were not the same issues that other southern regions faced. While the rest of the nation grappled with the "negro problem," Stone observed, in the Delta "we hear nothing about an ignorant mass of negroes dragging the white man down." "We have but one negro problem," he claimed: "how to secure more negroes."[28]

Delta planters' obsession with an adequate labor supply was not unfounded. Historically, a great number of sharecroppers and renters in the cotton South moved at the end of every season, but in the three years leading up to the weevil's 1909 entry into the Delta, there was a remarkable new influx of workers. Ahead of the encroaching pest, thousands of cotton laborers moved from weevil-plagued regions to the Delta in an effort to escape the insect's damage. One Greenville paper identified an "EXODUS OF NEGROES" from the boll weevil territories. In southern Mississippi, where the pest was causing significant damage, "the negroes refuse to listen to the appeals of the [local] planters," and as a result "2,000 negroes have moved . . . into the delta." The Greenwood *Commonwealth* offered its own "Advice to Negro Tenants" who were already in the Delta: "Stay right where you are." The paper observed that the boll weevil had pushed workers into the region, and it prophesied just what planters feared most, that it would soon push labor out. When the pest arrived, the paper predicted, "many no account, trifling niggers . . . will have to hike it, but the country will be better off without them." Few planters agreed.[29]

Stone took the lead in making sure that the Delta was prepared for the coming boll weevil. In 1904, three years before the pest first appeared in the Magnolia State, he lobbied the Mississippi legislature to create a new branch of the state's experiment farm for the Delta. Stone and his fellow planters believed research done at the state extension service's other branches, especially its major research area at the Mississippi A&M campus in the eastern part of the state, was unreliable because of the unique soil, larger farm size, and other environmental, economic, and social factors unique to the Delta. That spring, Mississippi's governor signed into law an act to support a two-hundred-acre research farm in the Delta, but private funds would have to be found to purchase the land. It took Stone mere weeks to raise the money and find suitable acreage for the state farm. Not surprisingly, "a group of progressive Washington County landowners," as they were later called, decided on a two-hundred-acre tract abutting Alfred Stone's own Dunleith plantation in a settlement called Stoneville. When Stone and his neighbors produced $15,000 in cash to purchase the land, the state allocated $3,000 to the MDA to operate the new Delta Branch Experiment Station (DBES) for it first season.[30]

From the beginning of the Delta farm, the expectation of Stone and planters like him conflicted with those of the MDA. Stone wanted to create a place where experts would conduct research that would directly aid the work of his plantation, namely cotton production on a huge scale. The state, however, saw agricultural experimentation not as a means to bigger cotton crops on plantations, but simply as a way to better understand farming as a whole through observation and analysis of soils, plants, fertilizers, and seeds. With federal support offered through the Hatch and Morrill Acts, which had created experiment stations and land grant colleges like Mississippi A&M, the state of Mississippi was legally bound to make research findings broadly available to its citizens.[31]

In reality, however, relatively few small farmers in the state ever traveled to the experiment farms to see demonstrations, only a handful of Mississippi growers read the bulletins published periodically by the stations, and even fewer put any of this research into practice. As a result, though insect and crop investigations at the Stoneville station were publicly supported, in practice the farm was a private space open only to those nearby who could take the time to visit, or those who had operations large enough to employ a person to learn about the station's research. Closed off from the experiments and their findings were thousands of the Delta's sharecroppers. These tenants might have benefited from the research findings of the DBES experiments since they were the ones doing the actual cultivation, but Jim Crow customs kept black farmers from this advice. As the insect approached, planters in the Delta wanted to keep themselves informed about the coming threat but keep the labor force in the dark.[32]

In the four years of work done at DBES between the time of its creation in 1904 and the arrival of the boll weevil to the Delta's southern tip in late 1908, scientists generated no new plans for protecting the region's cotton. Within the USDA at this time there was a feud between proponents of fighting the weevil with increasingly powerful insecticides and those who favored a cultural method to limit weevil populations. LeRoy Percy had even had a hand in selecting the Stoneville's station's chief scientist with this feud in mind. He wrote of trying to find a man who could "placate these departments" that disagreed. The selection of Jesse W. Fox as the station's first director may have made peace within the warring USDA camps, but it did nothing to generate new recommendations.[33]

Despite the real environmental distinctiveness of the Delta, the advice to planters that emerged from Stoneville was the same as that doled out by farm experts across the South: grow less cotton. Cotton was the "greatest staple money crop that can be grown," DBES director Fox wrote in his 1907 report, but the "one crop system . . . is wrong both in principle and practice." Fox recommended instead that Delta farmers plant cotton only "as the principal *surplus* money crop." Landowners should grow the staple only "after the farm has been made self-sustaining." Fox's vision of a diversified Delta where planters first raised food for themselves and their workers and grew cotton only on the land left over flew in the face of landowners' agricultural ethos.

After all, they believed they had carved this cotton wonderland out of worthless, soggy marsh for the express intent of growing cotton, and they had been successful doing so.[34]

For planters, Fox's advice was not only unfathomable; it was clearly not what they had expected when they funded the DBES. Stone, Percy, and their neighbors believed that while diversification was a fine suggestion for southerners living on inferior lands, their Delta was the best cotton land in the world and that it could naturally withstand the weevil. The impediments that throttled cotton growing in the rest of the South were wholly absent from their region, they believed.[35]

This resistance to diversification was based on more than a feeling about the environment, however. Collectively, Delta planters had invested millions of dollars over decades to transform the landscape into one designed for industrial cotton production. In addition to labor costs, they had built gins, warehouses, rail lines, and other infrastructure necessary for cotton production, but applicable to few other kinds of agriculture. The plant also had a cultural grip on the whole of Delta society that is hard to overestimate. David Cohn defended planters' resistance to diversification with blunt simplicity. Even if a planter could "sharply increase his profits" by growing asparagus, he wrote, the farmer would refuse, because, simply, "asparagus is not cotton." "Arguments for diversification," Cohn concluded, "left men's hearts untouched."[36]

Diversification was not an option, so as quickly as they had raised the cash to build the DBES, the Delta elite backed away from any public interest in government help. They sought new ways of protecting themselves from the approaching pest. These newer solutions prized their own expertise and place atop the region's social and economic structure. In other words, planters sought Delta-specific solutions that bolstered their control over the land and the people who worked it. This allowed them to more closely control this scientific information and the manner in which it spread.

Two events encapsulate this shift in strategy. In 1908, when the first weevil was found in the southern Delta, the Greenwood Business League, a group of white business leaders, merchants, and bankers, developed a program of boll weevil education that stood apart from state efforts. In August, for instance, the league organized a night of lectures on "all subjects of interest to Delta planters." The speakers included Fox and E. R. Lloyd, also a researcher at the DBES. They spoke to the planters about the boll weevil and other farm topics and answered their questions.[37]

In one sense, the meeting was nothing out of the ordinary; planters were meeting with government agents. The difference was the setting. Although the Greenwood *Commonwealth* reported on the lectures after the fact, attendance had been strictly by invitation only. The league had sent announcements only to a small, hand-selected group of planters in and around Greenwood. In October, F. W. Sterling, secretary of the league, sent a list of these planters to the USDA, asking that the department send any and all material relating to the boll weevil to these large landholders. Clearly there was an attempt here to educate, but to rigorously limit access to knowledge about the pest. These efforts were more symbolic than practical, however. Any farmer in the state could write to the USDA and request the same information that the league was giving to planters, although the region's elite may have asked certain questions of Fox and Lloyd at the private meeting that they would not have asked in public, especially questions concerning labor management.[38]

The second event that represents planters' privatization and localization of anti-weevil strategies is the incident with the educational train in Greenwood in 1909 that opens this chapter. Planters and business interests were not opposed to hearing government experts talk about the boll weevil, but they refused to let this happen publicly because they feared its effect on their labor supply. To them, controlling the boll weevil meant controlling whether and how workers thought about it. Their actions in Greenwood speak to the power of the boll weevil myth—they knew that the region's black workforce had heard of the pest's power to destroy plantation agriculture and they wanted to stop that idea's spread, or to at least change how workers thought about it.

By managing access to information about the pest, planters were trying to change the narrative of the boll weevil. In the fall of 1910, two of the region's biggest growers embarked on their own course of research that sought to literally rewrite the pest's story. Alfred Stone, the planter integral in the founding of the Delta research station, and his neighbor, planter Julian Fort, made a fifteen-day trip through the infested territories of Texas, Louisiana, and Mississippi. The men rode from town to town speaking with "planters, managers, and negro tenants, merchants, cotton factors, bankers" and others. The point of the trip, the planters argued, was "for our own guidance in framing a policy for our planting operations when the long threatened boll weevil invasion shall have become a reality." Upon their return, Stone and Fort wrote up their conclusions, and the First National Bank of Greenville offered to fund the paper's publication and distribution. The result was a thirty-three-page report with the authoritative title "The Truth About the Boll Weevil."[39]

The pamphlet made no attempt to condense current research findings. In fact, the authors' distrust of USDA and MDA publications was implicit in the booklet itself. Stone and Fort opened their report with a blasting critique of the USDA's work to date, citing the "confusing, contradictory and hence frequently misleading character of the discussions." Stone and Fort argued that the research services overstated the pest's wholesale ability to destroy cotton cultivation once and for all, and that these predictions that the pest will "break up the plantation system . . . . and bring everything to the level of the small farm, with cotton as a surplus crop" are "still no more than predictions." Instead of this dire outlook, the planters offered an alternative vision.[40]

Stone and Fort drew the conclusion that the Delta could be saved from the weevil by its geographic location and the presence of its powerful planter class. The authors argued that the region was environmentally unique; it was unlike any that the pest had yet entered and this gave cause to believe that the Delta might be immune to the pest's destruction. The more important factor, Stone and Fort argued, was that in the places where the boll weevil had done its greatest damage there had been an absence of a powerful planter class. Stone and Fort bemoaned the panic of weak-willed planters and merchants in other localities. Elites in Texas and Louisiana had assumed the invasion meant "an inevitable and hopeless economic wreck." Panic, they argued, had left even the biggest farmers unwilling to even attempt to make cotton in the presence of the bug. In Louisiana, the authors "saw abandoned property, with idle gins and empty cabins." These places suffered from the "disadvantage" of having smaller farms and a diminished planter presence than the Delta. They criticized the failure of local leaders to "allay fright, quiet labor and instill confidence," and to "stand up and make a fight." That, Stone and Fort assured their readers, would not be a problem in their Delta.[41]

To make their point about labor even more clearly, Stone and Fort detailed an encounter with a human manifestation of this "boll weevil panic." North of Natchez they found "the country roads . . . were filled with negroes, wagons and mules." These migrant farmers appeared to be "puzzled . . . upset, disturbed and bewildered." From Stone and Fort's perspective, these black families were moving from a wasted landscape with a hopeless future to the greatest cotton region in the world. "The Delta planter appeared on the scene," they wrote of their neighbors and colleagues who had arrived to take advantage of the weevil's impact on Natchez. When Stone and Fort came across them, other Delta planters had already paid the tenants' local accounts and had begun transporting these "hundreds of negroes" and their "household plunder" north to their massive plantations. With the labor exodus from the boll weevil territories of southern Mississippi to the Delta in full swing, Stone and Fort predicted Natchez's immediate demise not only as a cotton-growing region, but also as a profitable farming region of any kind. By 1910 it had already become home to "wasted fields" and "dwindling crops," a "desolation" not seen since "the devastation of the Civil War."[42]

Stone and Fort interpreted their encounter with these migrating workers as a positive statement about the promise of the Delta, but the subtext was one of fear. If workers could so easily leave one cotton-growing region for another, they wondered, what would keep them in the Delta once the pernicious beetle arrived there? Their answer was to worry first about keeping workers on their land, and to think about the weevil only secondarily. Near the end of "The Truth About the Boll Weevil" they put a fine point on this argument. "We did not talk to a planter who failed to dwell on the fact that his damage [from the weevil] was in proportion to his ability to hold and take care of his labor," they wrote. "We cannot make cotton without labor," the planters railed, "and we cannot hold our labor if we pursue the suicidal policy of not only becoming frightened ourselves, but of showing our fright to our negroes." Paying attention to the fear, what they called the "bugaboo aspects of the boll weevil problem," was just as important as the insect itself. "Our conduct," the planters warned other large landowners, "will be reflected in that of our labor."[43]

Only incidentally did the report address practical techniques of fighting the boll weevil. There was no discussion of crop spacing, pesticides, or fertilizers. And though Stone and Fort assured their readers that diversification was a plausible solution for other parts of the South, there was "danger" in applying that advice to the Delta. The authors reluctantly encouraged farmers to plant crops other than cotton, but only on "surplus land." The pamphlet stood in direct opposition to the advice being given concurrently by most farm educators across the South. Instead of advising farmers to plant other cash crops and home supplies (fruits, vegetable, food for animals) as federal and state agents recommended, Fort and Stone put the emphasis on cotton, and relied on the land and the ingenuity of their class to overcome the insect pest. Planters who would be most successful against the boll weevil, the authors argued seemingly without irony, would be those who "stuck to the crop with which they were accustomed to grow," despite the arrival of a creature whose life literally depended on the destruction of that crop. The sum of their advice for their fellow planters who read "The Truth About the Boll Weevil" was this: don't panic and plant more cotton.[44]

As Stone and Fort's pamphlet shows, Delta landowners' perception of sharecroppers—their deep-seated white supremacy based on a "natural" understanding of racial hierarchies—shaped their treatment of the landscape, which in turn affected their employment of black labor. Planters endeavored to keep growing cotton because in the 1800s men and women in their employ had transformed their land into a cotton-producing Shangri-La. They had directed black labor to create this place and they would fight to keep African American workers there to protect it from the boll weevil. As "The Truth About the Boll Weevil" argues, only the plantation system of agriculture implemented by "fearless" men like Percy, Stone, and Fort on *this* land could fend off threats from nature.

If and when LeRoy Percy read "The Truth About the Boll Weevil," it is certain that he would have agreed with Fort and Stone that beating the boll weevil was a labor issue rather than an agricultural one. But in his writings, Percy argued that the planter class was part of the problem, not the solution. On the eve of the weevil's arrival in the Delta, Percy argued that on some area farms sharecroppers had too much independence from white oversight, and that as a result, planters would not be able to control the details of cotton planting that the weevil's coming would demand. In Percy's view, the beetle was sure to decimate cotton as a result of this lax oversight, which he attributed to new credit arrangements. "Without question the weevil will bring with him disaster and pecuniary loss, due to the unprepared condition of the Delta," Percy wrote. "It is not prepared now for the weevil, and will not be when it reaches here . . . principally due to the fact that the present economic conditions in the Delta are fundamentally wrong." "Credit has been the curse of this section," he wrote, mainly because of its effect on black labor. Percy described a new mortgaged planter class as "men without any experience, ability and pecuniary resources." The result was a group of landowners who didn't have a history of managing laborers. "Easy credit has brought about the demoralization and deterioration in the negro labor of the country," Percy concluded. No matter the cause, to Percy the result was that sharecroppers had too much independence and agency in cotton production.[45]

It is not surprising that Percy's, Stone's, and Fort's perceptions of their largely black workforce were rooted in their white supremacy, but the way that that racism manifest itself in the face of the boll weevil threat is revealing. Underlying these men's agreement that the key to controlling the boll weevil was managing laborers' access to information was an assumption that planters controlled sharecropper life. Embodied in this attempt to control the boll weevil narrative was planters' racist notion that "labor may be kept from the false ideas and impressions which the ignorant easily gather," as one observer put it.[46]

Area newspapers too questioned the capacity of African American farmers to understand the reality of the boll weevil threat. "The average negro has never done anything in his life but raise cotton, and under favorable conditions he is good for that, but for nothing else worth mentioning," wrote the Greenville *Daily Times*, "but he balks at the boll weevil." The insect "is something [the sharecropper] cannot understand," the paper explained; "he simply knows that cotton will not mature where the boll weevil exists, and he is running away from him." That final point is revealing; planters and other white Deltans saw tenants' emigration in the face of the boll weevil as evidence of simple-minded fright, rather than a realistic, even technically sophisticated, understanding of the pest by farmers who worked that soil and that cotton crop year after year. Yes, tenants understood that the bug was geared to destroy cotton growing in the entire region, but evidence suggests this understanding came from experience, not fright.[47]

*

Delta writer David Cohn wrote that in his home region, "the Negro, for his part, must work out his destiny within a framework created and ordained for him by the white man. He must be all things to all people, an actor who never steps out of character." In reality, however, tenant farmers rarely read from the script written by white elites.[48] Just as planters sought to reshape the boll weevil myth in the Delta, black Mississippians also sought to use the idea of the pest's transformative power to improve their own lots. As environmental historian Mart A. Stewart has explained, black southerners had long used the natural world as a source of resistance to those in power, and the case of the boll weevil was no exception.[49] One episode in particular, explained in an extended correspondence between LeRoy Percy, his son William Alexander Percy, and a client, provides a telling story of how tenants worked within a racist and economically oppressive situation to use the boll weevil myth as a means to climb out of poverty.

In October 1908, LeRoy Percy, as he did most autumns, found himself concerned about emigrating tenant farmers. Of particular concern was L. A. Saunders, a white renter who had skipped town for Arkansas. Saunders was not the typical emigrant, however. He had rented an entire plantation from Johanna Reiser, a widowed client of Percy's who lived in New York City. Late in the 1908 season, Saunders ran off and Reiser wrote Percy asking for his help. The planter and lawyer told the widow that it would be easy to sue Saunders for the rent due, but that it would be much harder to actually collect. No one was quite sure where Saunders had gone. The still bigger problem for Reiser, Percy explained, was "endeavoring to secure a tenant in his stead."[50]

Reiser was in trouble. Percy assured the widow that finding a tenant to rent the land at this late stage in the season would be impossible and that her prospects for renting or even selling the land the following year were slim. "The outlook is intensely depressing," Percy wrote; "there is great alarm felt about the approach of the boll weevil, so great that it is practically impossible to make a sale or lease of property in this section. It is estimated that the weevil will be here and do considerable damage probably next year, and great damage after that." The beetle was disrupting the demand for, and prices of, land. Percy wrote the owner of the plantation neighboring Reiser's, asking if he might rent the land, but was rejected. That planter also believed that farming more land during the weevil's invasion was no advantage.[51]

Out on Reiser's plantation, the sharecroppers that had contracted with Saunders realized they were in trouble too, and tried to engineer a way to take advantage of the situation within the limited realms of power open to landless black Mississippians. They, like Reiser, had no recourse against Saunders, and were unsure of their future. They did know, however, that Percy represented Reiser in the matter, and so a couple of the tenants rode into Greenville and knocked on the door of Percy's office. They informed the lawyer that they were under no compulsion to stay on Reiser's land, and that if there was no one to furnish seed and supplies to them for the following season, they would leave that fall for a different arrangement somewhere else. Tenants were almost never in the position to make demands or threats of landowners (or their powerful representatives) and this was no exception, but these tenants were at the very least expressing a simple fact: they would take their labor elsewhere if the situation was not soon resolved.[52]

In New York, Reiser, after receiving Percy's dire appraisal of the situation, asked Percy to take over the land himself, as a favor to her. "You ask that I take hold of the property as if it were my own," Percy responded; "this, of course, would involve a very considerable outlay, with the result doubtful." If the land was his, he assured her, he would have already hired a manager, and secured and furnished labor, but that this expenditure was risky with the boll weevil's arrival. Percy had his own land to worry about and doubted that anyone else would take over his client's farm. "The apprehension regarding the boll-weevil [*sic*] is so great that no one will undertake to handle the property," he wrote. But perhaps in light of his recent visitors there was one solution.[53]

Percy suggested taking advantage of the one group in the Delta who would jump at the opportunity to rent her land: the sharecroppers already living there. "There are some negroes upon the place that have been apparently good tenants, and hate to leave," Percy wrote Reiser, "and I am satisfied that a lease of the land can be made to some of them." Percy suggested renting the land "at almost any price to any any [*sic*] negroes who are able to secure advances." If these tenants could find financial backing from a merchant for seed and supplies, any rent that they could pay Reiser at the end of the season would be profit, certainly more money than what they would most likely get from the delinquent Saunders.[54]

The problem with this plan, however, gets to the heart of the South's credit dilemma, a situation exacerbated by the boll weevil. Just as no person with sufficient capital would rent Reiser's land because of the pest, merchants were extremely cautious about advancing credit to farmers attempting to make a crop in the presence of the bug. If the farm lay fallow for the year, however, the land would suffer even greater damage, making it still more difficult the following season to find a renter.

Percy suggested a way around this conundrum. "If you were in a position to advance these negroes in cash, through the Bank here, as much as fifty cents per acre per month," he wrote, "I expect that all of them who have not left the place could be kept on it." Passing the financial risk for the crop onto Reiser, Percy was attempting to limit the influence of the Greenville community's fear of renting directly to black farmers and of its reticence to extend credit in the face of the advancing insect. Percy admitted to Reiser that "it is a dangerous thing, situated as you are, to make these advances, but the alternative, provided you have the money with which to make it, is worse." If she let the land lay fallow for the year, the natural decline of the fields, Percy argued, would mean a drastic depreciation in the land's value. He wrote that he could "do nothing for the negroes, and there is no doubt about the fact that it would be rather a risk for anybody to advance them." He admitted finally that, in all likelihood, "little or none of [the land] will be cultivated" in 1909.[55]

A gap in this exchange of letters between January and November 1909, leaves the fate of the land during that season unknown, but it offers a point at which it may be instructive to analyze why this exchange is so revealing. Percy, one of the South's most powerful men, a personal friend of Theodore Roosevelt, was dealing with an absentee landlord based hundreds of miles from her Mississippi plantation, and he was trying to persuade her to use her own personal wealth as a kind of credit buffer against the fear of the boll weevil. This case is by no means typical, but the length that everyone involved had to go to even attempt a labor-landowner contract that broke from the norm, to rent the land directly to the black sharecroppers already living on Reiser's plantation without a local overseer, underscores both the power of people's fear of the boll weevil and the basic unwillingness of creditors and landowners to give tenants the opportunity to gain economic freedom. It seems from the extant letters that these extraordinary forces were actually overcome and that the sharecroppers became renters that year, but Percy and Reiser had not settled their problems with labor, the boll weevil, and the land.

Letters from November 1909 and after suggest not only that Reiser was successful in renting her land directly to the laborers the previous season, but that the sharecroppers-turned-renters grew a bumper crop. That fall, she made a trip to Greenville to check on her plantation and oversee some general improvements to the property. While there, Percy sent her "some blank rental contracts and rent notes," along with an instruction that she fill out the contracts with the tenants and distribute the documents among them. After making the arrangements, Reiser returned north, and the tenants were left to farm the land in the 1910 season as they wished. Percy makes no mention in this exchange of ever going to see the tenants or instructing them in any way. In fact, when he reported to Reiser in June 1910, he seemed surprised to admit that "the crops are in very good shape." He based this conclusion on the account of one of the tenants, "the old negro who is sort of the head of affairs," who had stopped into Percy's office to let him know the condition of the cotton. The tenant, who Percy described only as the man wearing a "long plat of hair across his forehead," assured the lawyer that all but seventy acres of the plantation were planted in cotton and that its condition was excellent. "The old negro seemed to be sincerely pleased with himself," Percy wrote.[56]

Though no records exist showing how much money each tenant cleared at the end of the season, it is apparent that all but one made enough to pay Reiser the rent she was due. In November, Percy's firm wrote to Will Howe, one of Reiser's renters, that he was "the only renter on Mrs. Reiser's place who has not paid up and I do not intend to let you off with any of the rent. If the rest can pay you certainly can." The evidence that sharecroppers had a second successful year now in the presence of the boll weevil says a couple of things about the opportunity and risk that arrived with the pest. First, the tenants had raised two successful crops, the second one coming while neighboring plantations struggled with significant weevil damage. Second, despite this success, the tenants' material conditions had not improved. In spite of reduced land prices and two years of profit, these tenants were still in no position to purchase land. The renters' incremental advances up the mythical farm labor ladder still offered no lasting security. Only landownership could provide that safety net, and the men and women on Reiser's land had little hope of raising enough capital or acquiring the necessary loan to purchase the land that they had saved.[57]

From the tenants' perspective, the boll weevil had presented a rare opportunity. The freedom to farm without white landowner direction meant agricultural, economic, and social freedom rolled into one. It meant independence, or at least a kind of localized independence. This chance at moving up the labor ladder from sharecropper to renter would last, the correspondence suggests, for at least one more year.

Despite the apparent success of the endeavor, Reiser still looked to sell the land during the 1910 season. At this point, LeRoy Percy was joined in his practice by his son William Alexander Percy, who wrote to Reiser that there was little hope of selling her property that fall. "The boll weevil is expected in small quantities this years in Washington county [*sic*]," the younger Percy wrote in 1910. "All of the planters," he claimed, were "very blue" and "disconsolate." Land prices in the Delta had plummeted as the pest crept into the region. LeRoy himself had turned down the purchase of a large plantation that his firm had managed "because of the approaching boll weevil." The plantation eventually sold, Percy told Reiser, at a "ridiculous low price." Unable to sell her land, Reiser returned to the business of absentee landlordism.[58]

In 1911, despite the black renters' two years of success, the Percys still recommended that Reiser find "a responsible white tenant" to take over management of the plantation for the upcoming season. "You have every right to be pleased with your *experiment* this year," William wrote, "but I am free to confess that I never expected you to collect one-half cent rent due." Despite their achievements, if Reiser continued to rent to the black tenants, they would undoubtedly still need a white person to advise them, to manage the land, and to oversee their efforts. "As long as negro tenants occupy the place, without making repeated trips to the plantation itself, which we cannot do," the younger Percy warned, "it is quite impossible for us to adequately protect your interest."[59]

Reiser was not convinced. She realized that her advisors' suggestion came with both cost and risk. A white renter would demand more of Reiser as a landlord than her black tenants had. Even if she could find a white renter, which she doubted, "he would require repairs which I do not care to make and the result would be, I would lose my present colored tenants, and the white man would never turn up." The delinquent Saunders, it seems, had indirectly taught her a lesson. "About my experience of a white man honestly?" she explained, "well no colored man could or would have cheated a widow worse than I have been taken advantage of." Reiser appreciated that the sharecroppers-turned-renters had stayed with her land and made a successful crop; she declared "I will again rent to colored tenants."[60]

By the autumn of 1910, the tenants had circumscribed the Percys in their own communication, writing instead directly to Reiser and making arrangements for the upcoming season. "They all write they will be glad to stay at the same rate [of] $5.00 per acre," Reiser wrote Percy, "each tenant is his own boss on his own land for which they pay rent. You please tell them so also." Reiser stood by the men and women living on her plantation. The abandonment of the land by the white renter along with the fear generated by the boll weevil had, for a brief moment, overcome the economic and cultural power of the tenant system.[61]

As the Percys continued to try to find possible buyers for the Reiser place, this same group of black Mississippians remained working, appreciative of a landlord who would never visit or give unwanted advice. The correspondence trails off in the spring of 1911 and there is no record of the fate of Reiser's renters that year. It is clear from the exchange, however, that the boll weevil had a great influence, if only for a moment, on the financial prospects and social freedom of a group of black Delta tenants. Though admittedly rare, this example offers a glimpse of the power of the boll weevil's threat to the Delta at large and to planters in particular. Though generally the arrival of the boll weevil meant landowners tightened their control over labor, in cases such as this, their fear of putting in a cotton crop opened up space for tenants to seize increased control over their work lives.[62]

By 1911, the tone of the Percys' letters to Reiser began to change, suggesting that these Delta patriarchs were learning something from her experiment in plantation management. The younger Percy reported to Reiser that "your place is now all rented, and as far as looks go the tenants seem capable and well to do." The result for people like the Percys was a degree of surprise. Though planters had been obsessed with the labor force in the Delta, constantly deriding their poorer neighbors while simultaneously feeling that they were themselves proper paternalists helping out this "shiftless" workforce, the success of the renters on Reiser's farm opened their eyes to alternative notions. As the tenants planted cottonseed in March 1911, the younger Percy wrote to Reiser admitting that the boll weevil and this group of black tenants had taught him something. "The experience has been an excellent one in human nature," he wrote, "and has given me considerable insight into the methods and characters of the darkies."[63]

On the surface, Percy's admission is loaded with the racism, paternalism, and condescension that lay at the root of most rural white Mississippian's relationships with African Americans in the early twentieth-century Delta. Though he admits to have gained "considerable insight," the language suggests that his feelings about black farmers' capabilities would not be overcome simply by his personal involvement with them, no matter the outcome. This experiment was the exception that proves the rule, Percy seems to have been suggesting; in general, black Mississippians were incapable of learning to fight the boll weevil without white supervision. But on a deeper level, it is important to recognize just what Percy is admitting to. The demeaning language aside, the boll weevil had indirectly taught a lesson about the capabilities of black tenants to one of the South's most powerful families, and it was a lesson that the younger Percy at least admitted having learned. It remained to be seen, of course, just how lasting this lesson would be.

*

The myth of the boll weevil's wholesale destruction of the plantation system had disturbed Delta society even before the pest itself reached the alluvial region. Planters, tenants, educators, and researchers jockeyed for power as the pest approached; each group hoped to seize on the weevil's arrival to either change their positions in society or further entrench them. The fear of the weevil shook up the Delta; it rattled planters' control over the land and the people and opened up spaces for other groups to gain power.

State and federal scientists tried to dovetail their recommendations for the unique environmental, social, and economic conditions of the Delta. From demonstration farms, lecture halls and train stations, these educators preached diversification—less reliance on king cotton—for some audiences, and taught industrialized cotton production to others. In so doing, they tried to avoid stepping on the toes of those who wielded supreme economic power in the Delta, the planters and merchants.

These elite men and women were also working to grapple with the changes the boll weevil would bring. Concerned first and foremost with securing their labor force, planters tried with some success to control the movement of the extension agents, of information about the weevil, and of the pest itself. The boll weevil exacerbated planters' misconceptions of their control of their labor force. In managing information, the weevil myth itself, planters believed that they could control their labor and ensure the future of Delta cotton.

For tenants, the boll weevil had already proven its power. It was the pest that had pushed many Delta sharecroppers off cotton fields in Texas, Louisiana, Oklahoma, and Arkansas. Many knew firsthand the destruction the tiny bug could cause. In the Delta, they found a community of black labor like nowhere else, and as the boll weevil approached and destroyed its first cotton bolls there, sharecroppers continued to move around within the region. But from 1908 to 1912, during the weevil's initial foray into northwest Mississippi, there was no mass exodus of labor from the Delta.

By 1913, the boll weevil had been present in the southern Delta for three years, but still had not yet reached the Tennessee border to the north. It was the decade that followed this initial foray through the region that would decide the fate of planters and laborers.

CHAPTER 4

Delta Solutions Big and Small

The four Englishmen had not dressed for the weather of the Yazoo-Mississippi Delta. It was the spring of 1911, and they were ostensibly on an outing to hunt geese. But as they stood at the bottom of a specially dug pit on the east bank of the Mississippi River, they looked to observers to be hot and pathetic. They had taken off their jackets and unbuttoned their shirts. One local woman who witnessed the scene said later, "You could see their wool underwear. And they were just dying of heat and they looked like four bums." These "bums" comprised the board of directors of one of the largest mill conglomerates in the world, the Fine Cotton Spinners and Doublers' Association of Manchester. They had travelled from England to the Delta in search of cotton land, to ensure their mills' supply of the fiber. This excursion to the Delta coincided with the arrival of another visitor, the cotton boll weevil.[1]

When the Englishmen returned to Europe, they sent back word that they would purchase that land on which they had sweated and hunted. With its sheer size and vast capital resources, the Fine Spinners believed they could buy this huge tract of cheap land—fear of the encroaching weevil had depressed land prices—and guarantee themselves a never-ending supply of long-staple cotton. Despite the threat of the boll weevil, a force that these Englishmen believed they understood, the company's board of directors envisioned a Mississippi colony, an industrial farm producing the fleecy fiber on a massive, automated scale. The boll weevil trampled *that* dream, but it allowed the rise of an alternative one.

After a series of end runs around state law, they formed what was at one time the single largest cotton plantation on Earth. The Delta and Pine Land Company (DPLC), as it would eventually be known, farmed cotton on a massive scale and employed dozens of the South's top farm scientists, hundreds of white managers, and thousands of black field workers. For most of the twentieth century, its presence in the Delta was impossible to miss and its legacy is hard to ignore. It remains to this day one of the South's biggest and most influential agribusinesses. Perhaps surprisingly, DPLC's fight against the boll weevil is connected to both its legacy and its success. In 1911, when the Fine Spinners took control of the land, the Delta simultaneously became home to the world's largest cotton farm and its greatest insect enemy.

When it destroyed any prospect of long-staple cotton production, the boll weevil created a new DPLC, a farm that would make its money by pioneering ways to fight the pest. In essence, the company cashed in on fear of the cotton insect by selling its own weevil-beating resources to farmers throughout the South who had been persuaded by the boll weevil myth that profitable cotton growing was no longer possible. DPLC offered an antidote. By luring the region's top scientists away from the public sector and spending millions of dollars on research, DPLC sold its ability to resist the boll weevil. It took advantage of the boll weevil myth, and by marketing its solutions to the pest it furthered the myth's spread and force.

DPLC's success in taking advantage of the boll weevil embodied a telling irony. Despite the modern industrialized farm's ability to fight the pest on a scale unlike any other in the South, and with natural and synthetic weapons few other farmers could get their hands on, its operation still relied on the tenant system to carry out the labor of cotton growing. Sharecroppers at DPLC found themselves part of a corporate order lacking some of the traditional trappings of postbellum plantation paternalism. DPLC managers controlled every aspect of laborers' work lives and attempted to control even their social and personal spaces as well. Croppers lost the independence they had relished on other farms but carried a new burden, a kind of corporate control over their lives. From the tenants' perspective, everything from the increased regimentation of their day-to-day work lives to periodic exposures to experimental chemical insecticides made them a kind of Old World relic amidst a futuristic agribusiness.

Within this radical and contradictory sharecropper experience, black Mississippians sought their own modern solutions to life in the cotton fields. Thousands moved on from Mississippi, as so many had from the weevil-infested fields of Texas and Louisiana. But this time many left the rural South altogether for the promise of the industrialized North. Others expressed their discontent by making and consuming their own boll weevil stories. As DPLC built its modern operation, the first recordings of black Mississippians singing songs about the boll weevil made their way into the jukeboxes and dance halls of the nation. And though the socioeconomic power that this cultural expression gave Delta tenant farmers is limited, the music's legacy is important. These boll weevil songs amended and amplified in the Delta are some of the most durable cultural expressions of the entire twentieth century South (certainly more Americans today know Howlin' Wolf than the Delta and Pine Land Company). These boll weevil blues contain snapshots of how workers at once embraced and rejected the changes that the tiny beetle had brought to this New South plantation kingdom.

*

As the first boll weevils appeared in the Delta, planters and sharecroppers alike paid close attention to its exact location. The pest tended to travel quickly through areas where cotton was sparse, but slowed where the plant was most dense. Though the weevil had raced from the southern edge of the Mississippi-Louisiana border north to Vicksburg, it slowed precipitously when it reached the cotton-rich Delta. It had made Vicksburg at the end of 1908, but only crept north the following season. In the years that followed, the pest charged north along the river, but slouched through the most densely planted inland counties.

Even with the pest at their doorstep, many observers of the Delta scene could not agree on the potential threat that the weevil posed. Alfred Stone and Julian Fort had argued that the Delta's uniqueness—its land and labor—could resist the weevil's ravages, and the publication of their pamphlet "The Truth About the Boll Weevil" sought to convince their neighbors of the same. Planter Walter Sillers was swayed. He wrote in March 1909 that despite the fact that the beetle had not yet made its way onto his land, that "the boll weevil devil, like all other devils, dont [sic] seem to be as black as he is painted."[2]

The planters' conclusion was at odds with most agricultural experts. Two farm scientists argued in 1913 that:

It is utterly impossible for the farmer to make a crop of cotton with the boll weevil present under the old system of farming . . . [which] has been materially changed in every section where the weevil has yet appeared. The people have been forced to abandon the all-cotton system and to adopt the method that will enable them to produce all of the home supplies.

Like the planters, these state-funded researchers were attempting to seize on the boll weevil myth, to use the pest's threat to southern agriculture to bolster their own research and institutions. Though the authors promised that limited cotton production could be continued if it were part of a diversified modern farm, they claimed that industrial production in a mono-crop system was impossible.[3]

Despite the prognostications of cotton's demise, in the fall of 1909 the Delta's fields bloomed white. Steadily high cotton prices made a more powerful argument than even the most calamitous predictions of crop loss. A worldwide cotton shortage had pushed up prices, to the benefit of Delta planters, who increased their acreage even in the face of the approaching weevil.[4] On the demand side of the cotton equation, however, many of the world's cotton mills had begun to rethink their access to the fiber as prices had risen. Companies didn't care where the next load of cotton came from, as long as it arrived at a low price. Mills in the United States and Europe looked to guarantee their access to cheap supplies of the staple through vertical integration. The Delta's soil, climate, topography, and available land made the region a prime target for international investors seeking access to the fiber that would churn through their looms.[5]

In 1910, Jesse W. Fox, the newly appointed director of the Delta Experiment Station in Stoneville, delivered a paper in Brussels, Belgium, on "The Causes of the Present Shortage of American Cotton and the Means to Adopt to Prevent a Recurrence." Not surprisingly, he argued that the limits of American cotton production had not been reached, that in fact there were broad expanses of rich Mississippi land still available for planting. His talk caught the attention of the directors of the Fine Spinners and Doublers' Association. Representatives of this consortium of English mills asked Fox about the availability of land near his Stoneville base. They even went as far as suggesting that that if the company purchased land in the Delta, they would only do so if he agreed to manage their plantation.[6]

Fox returned to the United States and resumed his experiments. The professor was in no position to involve himself in a land deal with a gigantic English company like the Fine Spinners, but he may have mentioned his meeting with the mill to Charles Scott. Like most large landowners in the region, Scott was not merely a planter; he was a banker, a real estate developer, and a politician. Scott had in fact played a major role in the early development of railroads in the Delta—his own 11,000-acre plantation had a stop on the Illinois Central's line marked on maps as simply "Scott."[7]

Scott was a savvy investor and planter, and the threat of the boll weevil to his rural kingdom shaped his business strategy during the first few years of the pest's presence. In 1910, the first year of the weevil's appearance on his plantation, he saw a remarkable drop in land prices. Just as this decrease in land values had made it impossible for Johanna Reiser to unload her plantation that year, Scott recognized that the boll weevil panic was creating a major economic opportunity. Though most planters embraced a wait-and-see attitude when it came to the boll weevil (holding land they might otherwise consider selling and refraining from buying any new land), Scott took a gamble. Area land prices had plummeted as the result of experts' predictions that the weevil would end cotton production in the Delta, and Scott sought to take advantage of the depressed prices. In 1910 he exercised an option to buy 21,000 acres adjacent to his already vast property. He had quickly tripled his land holdings at a price he considered a value. Scott now had to decide whether to manage the vast land as a single plantation, growing cotton even in the face of the boll weevil threat, or finding someone to buy the entire lot. Scott probably knew that by buying relatively inexpensive land, he could sell the one large tract to an interested buyer from outside the region. He may have in fact known for certain from Jesse Fox that the Fine Spinners were interested in purchasing a Mississippi plantation.

At this point, Scott enlisted the help of Lant K. Salsbury. Salsbury was a young northern entrepreneur, a stark contrast to Scott, the archetype Old South planter-aristocrat (who had, in fact, ridden alongside Nathan Bedford Forrest as a young man during the Civil War). A Michigan native, Salsbury earned a law degree from the University of Michigan before working a stint as a lawyer in Grand Rapids, where he became involved in the purchase and sale of timber forests. Recognizing the demise of the timber industry in the Great Lakes region, he moved south at the turn of the century hoping to make his fortune in another extractive industry. Salsbury settled in Memphis and began managing a plantation in the northern tip of the Delta, south of the city. Salsbury was new to the South and had a vision of extracting its natural resources and using his northern connections and capital to buy and sell southern cotton and timber land. Scott embraced this outsider in an attempt to gird his own fortune and interests against the boll weevil.[8]

Salsbury assured Scott that, despite the pest, he could sell the planter's vast Mississippi lands at a substantial profit. With his connections and salesmanship, Salsbury, in fact, had designs on the land himself, though it is unclear whether he made that known to Scott. In either case, Salsbury joined with seven investors and purchased Scott's 33,000-acre tract overlapping Bolivar and Washington Counties for an undisclosed sum. Salsbury believed Scott's asking price was low, even considering the threat of the weevil, and attempted to turn right around and sell the land at a profit. Aware that few in Mississippi, or perhaps the entire cotton South, would purchase the land knowing of the approaching pest, he returned north, hoping to find a buyer. When no one in the Midwest or Northeast was interested, Salsbury took his sales pitch to England. He met there with representatives of the Fine Spinners, who had remembered Jesse Fox's promises of productivity in the alluvial Delta. They agreed to make a trip to Mississippi to see the land in person.

The Fine Spinners had been worried for decades about their supply of cotton. As a result, the mill owners looked for a way to streamline their supply chain from the cotton fields to their fifty English mills. The Fine Spinners were looking to guarantee their own access to the raw material and to limit the influence of speculators along the supply chain who, they believed, gouged the mills with high markups. In 1910,a group of the company's directors traveled to the Delta, where Salsbury and his partners took the overdressed Englishmen on that hot goose hunt. Locals took a keen interest in the foreigners' visit. "It is also rumored that the gentlemen who visited the property are interested in the cotton industry in England," the Memphis *Commercial Appeal* reported; "if this proves to be a fact, such a purchase will have the effect of turning the eyes of capitalists to the delta country."[9]

In the papers alongside news of the Englishmen's visit appeared stories of the boll weevil's arrival in the Delta. Newspapers, banks, and merchants both nationally and internationally had been keeping an eye on the advancing bug, and the Fine Spinners must have been paying attention as well. But the presence of the weevil had made the land cheaper, and in their view it must have been a price that the company believed was low enough to offset the risk. In April 1911, the Fine Spinners agreed to purchase the land for $3 million. Herbert Lee, the only member of the Fine Spinners' board to object to the deal later quipped, "They dug some goose pits and we fell into them."[10]

Delta newspapers heralded the land sale as the arrival of a progressive business interest in the Delta. The *Commercial Appeal* reported on DPLC's goal to build an entirely self-sufficient industrial farm. "The company intends to use improved methods of cultivation and to employ for this purpose an expert man as general superintendent," the paper wrote. DPLC indeed planned to build its own cottonseed mill, several gins, and compresses. "Every modern method known in the cultivation of cotton will be put in practice," the paper reported. The modern, efficient plantation envisioned, the *Commercial Appeal* promised, will produce "up to a bale and a half or two bales to the acre."[11]

DPLC took immediate possession of the land and assumed control over the buildings and materials on it, including the 1911 crop. The Fine Spinners hired Salsbury to manage the company. He oversaw the plantation's vast holdings from his office in Memphis. Behind a desk 140 miles from the cotton fields he was to preside over, Salsbury must have recognized the large task in front of him. Not only did DPLC own the equivalent of sixty square miles of cotton land, a scale on which no one in the United States had ever tried to grow cotton, but the boll weevil was knocking on the Delta's door.[12]

Salsbury knew that his own limited experience running a plantation left him unequipped to manage the day-to-day farming decisions of this mega-plantation. As a result, in his earliest days at his new position he made the first in a string of moves to use the land's size and the parent company's deep pockets to lure some of the most knowledgeable farm experts in the South to DPLC. Salsbury contacted Jesse Fox, the original director of the Delta Branch Experiment Station who had himself been hand-picked three years earlier by Alfred Holt Stone to run the Stoneville research farm. Salsbury courted Fox with an offer to leave his work at the state-supported farm, where his responsibilities were research and public teaching, for DPLC, a private company whose primary interest was profit.[13]

Fox had much to consider. He had been a pioneer of cotton research in the Delta since 1904, when the state, with the help of local planters, founded the Stoneville farm. In a short time, he had made remarkable improvements to the farm's 250 acres. Despite the rough condition of the Stoneville land, by 1906 he had managed to bring 210 acres under cultivation and started a series of experiments that by 1911 had already influenced the way many Delta farmers planted cotton. Fox tested various commercially available cottonseeds, investigating each variety and publishing his findings. Building on the work of the USDA in Texas and Louisiana, Fox attempted to identify those seed types that matured earliest, thereby limiting the amount of damage the boll weevil could do in the late summer and fall, and to breed these varieties to create new strains that took advantage of the best aspects of each type. Fox also researched immediately practicable aspects of raising cotton in the Delta, including factors like the timing of first planting, row width and distance between rows, soil preparation, cultivation, and fertilizers. He quickly built the Stoneville farm into one of the state's most productive branches and made a name for himself by presenting his research findings across the state and around the world. In addition to his work on the farm, Fox had been a staunch supporter of public education. He had taken his findings to the farmers of the Delta—in fact, he was one the experts on the Southern Railway's train that planters had refused to allow to stop in Greenwood in 1909. Less than two years later, Fox was considering abandoning his work with the state in favor of a commercial enterprise unlike any other in the South.[14]

The crux of Fox's decision whether to leave the public sector for DPLC lay not only in his personal research, but also in the ideological problems inherent in educating farmers of his findings. The region's commitment to cotton was more than simply an economic or agricultural decision to plant the staple; it was an ideology. Cotton was woven into the fabric of the Delta culture, which meant local cotton planters had a say in Fox's research. Despite his work in Stoneville on alternatives to cotton, the farm's location within the Delta and the power that area planters had over his work relegated Fox to spending the bulk of his time and resources with that staple crop. As he explained in his 1908 research report, "The important relation that cotton bears to Delta farming can hardly be overestimated . . . it is the greatest staple money crop that can be grown in any section of our country." But the region's commitment to the staple, Fox realized, was a dangerous thing, for it "has led to our one crop system, which, of course, is wrong both in principle and practice." But what more could he say? Pushing for diversification in the Delta may have been in the general public's best interest, but it was not in the interest of the planters who had hired him, bought and built his research station, and constantly looked over his shoulder at his experiments. In fact, over the course of his time in Stoneville, Fox had gradually changed his advice as he came to understand planters' inflexible interests. He had begun encouraging planters to make their farms self-sustaining first, then to plant whatever land was left in cotton, rather than promoting a plan of outright diversification. "We do not advocate the exclusion of cotton from our cropping system," he assured readers of his 1908 report, "nor do we fear that any one thing, nor a combination of circumstances, will ever cause [cotton] to become an unprofitable crop." Fox weighed these factors as he thought about DPLC's job offer. What must have played a decisive factor in his thinking was that DPLC was not a typical plantation led by a single man. It was a new model, a corporate-owned farm that had no interest in sustaining itself with diversification. It was a business geared to produce cotton and nothing else.[15]

Fox took the job at DPLC. In 1912, he moved from Stoneville to the new company's headquarters in nearby Scott and became its first general manager with a salary of $7,500 per year. He immediately began to guide the massive company through a period of rapid hiring and administered the purchase of fertilizer, tools, and mules. He broke the massive 30,000-acre plantation into sub-farms and hired managers to oversee each section. He also began the search for the huge workforce needed both to put in the 1912 crop and to bring as much land as possible under cultivation. Over the first few years, the Fine Spinners supported Fox's efforts by allocating an additional $1.5 million to improve conditions on the farm.[16]

A number of setbacks struck DPLC in their first three seasons. In 1912, there was major flooding throughout the Mississippi River Valley. A levee gave way north of Scott, flooding fields and destroying the bulk of the company's crops. The following spring, with water still sitting stagnant on most of the land, Salsbury ordered laborers to plant whatever land was above water. One resident described how "the men swam the mules across the bayous and planted the high spots." Despite these efforts, repairs to the levee were unfinished when water rose again in 1913, and busted through the bank to wash out the skimpy crop that workers had managed to plant.[17]

In 1914 there were no floods, but boll weevils made their first major appearance at DPLC. The pest increased in numbers consistently throughout the season before building to an enormous population by fall. Weevils destroyed most of the company's cotton that year. Despite the sums of money and vast resources poured into the operation, in its first two seasons DPLC had managed only meager crops. As historian Lawrence Nelson has written of the early years at DPLC, "The big plantation with the big plan produced a pathetically tiny crop."[18]

During the early years, boll weevils shaped just about every decision DPLC made. The Fine Spinners had bought the vast Mississippi acreage to produce long-staple Egyptian cotton. Apparently encouraged by Fox's Brussels paper delivered years earlier, the company seemed to believe that Egyptian cotton could be grown in the Delta, and indeed in some spots along the river farmers had been successful in growing long-staple varieties. The boll weevil made this impossible, however. The longer fibers took a full season to develop and the arrival of the weevil had in essence shortened the season by several weeks because its population was so great in late fall. Long-staple cotton bolls were only beginning to form at the point of the summer when the weevil's numbers reached full strength. DPLC tried an experiment with the plant in 1911, but the samples "grew ten feet tall and produced not a bloom." Even long-staple American Upland cotton, a different species of cotton plant than the Egyptian, failed to bloom until very late in the season. By that time, boll weevils could devour the long-staple squares.[19]

Fox and Salsbury broke the news to their bosses: the new land would not produce the cotton they desired. The Fine Spinners were more than let down in their cotton preference; long-staple cottons were in fact the only kind of cotton that their mills could spin. By 1913, only the second full season of DPLC's existence, it was clear that the boll weevil had put an end to the Fine Spinners' dream of an unending supply of cotton for their factories. In fact, no Mississippi cotton made by DPLC *ever* made its way to the Fine Spinners' mills.[20]

Despite these setbacks, the English company did not immediately pull up stakes from the Delta. In fact, the company's leaders recognized that the insect's presence offered an opportunity for DPLC to remake itself not as a supplier of Egyptian cotton to English mills, but as a supplier of boll weevil solutions and other agricultural remedies to southern farmers. This transformation did not happen overnight. At first, DPLC simply began selling their short-staple American cotton on the open market. But soon Fox and Salsbury realized that, with DPLC's size and access to capital investment, the plantation was uniquely suited to develop new ways to fight the boll weevil. Partly out of the necessity to make their operation profitable by finding a way to grow cotton in the pest's presence, DPLC's managers sought both novel ways to farm and new breeds of cotton itself. Not only did the company have the advantages necessary to develop ways to grow more cotton than its neighbors, it could sell these resources and methods to other farmers who were panicked by the prospects of cotton farming under boll weevil conditions.

DPLC's main weapon was science. Rather than devoting every inch of its land to cotton production for the market, as its planter neighbors did, DPLC allocated significant portions of its vast acreage to research and experimentation. Despite two decades of work on the boll weevil problem by southern scientists, there was no magical solution. The cultural method prescribed by the USDA in Texas as early as 1894 still formed the basis of farmers' techniques for limiting boll weevil damage. One DPLC researcher remembered that in the early years, the pest "would just, more or less, go unchecked until the population became so big that they just ate everything in sight." Fox recognized that DPLC was uniquely situated to engineer a better solution to the boll weevil than those offered by the USDA. He turned his attention to two that he thought were underutilized: plant breeding and poison.[21]

On both fronts the company's strategy hinged on its ability to attract more experts. In 1915, Fox enlisted the help of Early C. Ewing. Ewing, like Fox himself, had been educated and trained by the State of Mississippi, and had started his career with the state research service. He too turned away from his role as a public educator and moved to Scott to work for DPLC. Ewing had begun plant breeding work at the Starkville research station in 1910 after receiving a graduate degree from Cornell a year earlier. While at Cornell, Ewing had worked with a pioneer in the field of plant breeding and was on the fore-front of putting these new theories to practice with cotton. He ended his brief stint with the Mississippi Agricultural Extension Service when DPLC came calling, mainly because, as he recalled, "I could be happier and could accomplish more in a commercial environment than in an institution." His reasons were not "overly altruistic," he admitted. "I thought I might improve my financial status," he remembered. Still, Ewing would say later that DPLC's job offer was "a long shot gamble for me and certainly for the company."[22]

Fox had courted Ewing to DPLC chiefly due to his expertise with plant breeding, but when Ewing arrived in Scott he recognized that the plantation was a long way from being ready to conduct any major breeding research. It was remote, he later remembered, connected to Greenville only by an un-paved road that was treacherous after a good rain. He was surprised at the "primitive conditions" and relatively unorganized state of DPLC's operations. It was a sprawling plantation with a great number of buildings, but no electricity.[23]

Despite the "rough and crude" surroundings, Ewing began developing a plant breeding research plan. He tooled with cottonseed varieties in the hopes of developing a short-staple cotton that would fully develop early enough in the season to limit the effect of the weevil. He began selecting and breeding together the fastest-growing and most productive short-staple varieties, but the process was not as simple as producing a plant that matured quickly. New breeds also had to produce the long, strong fiber that the mills demanded. "Where the boll weevil is a constant menace," Ewing recalled, "productiveness, earliness and disease resistance" were the most important factors.[24]

Ewing had a head start in this important research. In his final days with the MAES research station in Stoneville, he had traveled to Texas to investigate early-maturing cotton in use in that state's battle against the boll weevil. Ewing was encouraged by a hybrid seed being grown called Express. The Express seed was developed by plant scientists in Texas from the Bohemian Big Boll variety, and was still not commercially available to farmers. In fact, local researchers in Texas had decided that Express was unfit for the soil of the Black Land Prairies and discarded the bulk of the seed, but not before Ewing got his hands on a bushel of it. He brought the Express seed back to Starkville and conducted a field test in 1911. The seed beat all competitors in both earliness and productiveness, asserting itself as the premier cotton variety for the boll-weevil-plagued Delta. Despite the success of Express seed, the station's report for 1912 noted that Mississippi farmers would have a hard time finding the seed for their own use: "Unfortunately there are no seed [*sic*] available of the 'Express' cotton, the variety that took first place, but the Delta Station and others will likely have a few seed for distribution in another year or two."[25]

Not all farmers, as it turned out, would have to wait that long. Early Ewing ferreted away a supply of Express seed when he moved from Starkville to Scott. Though even the Delta's large planters had no access to Express, Ewing had made sure that DPLC would have an immediate supply. He later referred to his taking the seed from the public farm to DPLC simply as "fortunate," but it would prove to have multimillion-dollar ramifications.[26]

Ewing's bushel of Express formed the basis of cotton-breeding experiments at the company for the next fifty years. Ewing and Fox bred Express with other varieties, building on its positive characteristics. Soon DPLC not only grew Express seed and its offspring for its own cotton production, but it marketed the seed to planters across the South. "These varieties went into production at once," Ewing wrote later, "and with a few others have since served as foundation material for the breeding of several prominent kinds which are the mainstay of the cotton industry in the South today." Ewing's son, who also became a researcher at DPLC, explained the value of his father's work with Express in even grander terms: "If it hadn't been for that early variety of cotton called Express that he brought in here, why [DPLC] would have been out of cotton production."[27]

For the first few years, Express derivatives were enormously popular with farmers. DPLC had come upon a way to take advantage of farmers' fear that the boll weevil would end cotton farming across the South. The company had engineered a cotton variety that could limit the pest's damage, and it marketed the seed to take advantage of southern growers' fear of the pest. By the 1950s, the company's sale of seed almost eclipsed its profits on cotton production itself. (DPLC's biggest seller in the twentieth century was the Deltapine seed variety, a direct descendant of the bushel of Express that Ewing had brought with him from Mississippi A&M, netting the company millions of dollars in sales.) DPLC used the myth of the boll weevil's out-and-out destruction of southern cotton to sell the seed to frightened farmers not only in the early twentieth century, but for decades after.[28]

DPLC's success selling seed led to other anti–boll weevil strategies. Poisoning the insect had not been a fundamental weapon against the insect since its discovery in Texas. In 1894, USDA research had found that Paris green, a potent copper arsenic compound, could kill as much as 30 percent of a weevil population, but the costs of the chemical and the difficulty of its application precluded the department from recommending its use. Still, the possibility of a poison solution for the boll weevil was very attractive to Delta planters and farm researchers alike, for both practical and psychological reasons. As opposed to the cultural method of boll weevil control, which involved the year-round work of improving all aspects of cotton farming from soil preparation through cultivation and harvest, a poison could simply be purchased and applied, yielding immediate, visible results. In addition to the ease that poison might bring to the boll weevil battle, many farmers saw it as a more modern solution than the cultural method. Those planters who already employed modern accounting practices and industrial labor supervision embraced poisons as an extension of up-to-date farm management.[29]

In addition to the high cost of Paris green, there were major problems with its application. One Mississippian remembered that the job of spreading the poison fell to tenants. "They went through the field," he wrote, "and had a long pole across the mule's neck with a sack of the arsenic on each end and they'd go along and shake it and they'd go down the middles of the cotton, you see, and in that way they dusted two rows at a time." Farmers hoped the dust settled onto the plant buds and that weevils would ingest the poison as they bored into the plants. More often than not, however, the majority of the Paris green fell straight to the ground, or was blown away in a breeze. Even in the best conditions, there were questions about Paris green's effectiveness. There was no doubt that it would kill the pests, but if applied too heavily, it damaged the plant and reduced yield to a greater degree than weevils would have on their own. With costs prohibitively high for all but the wealthiest planters, Paris green was not an attractive option for most farmers. Still, throughout the South, thousands of people tried to make Paris green the answer to the boll weevil invasion.[30]

In 1909, Wilmon Newell of the Louisiana Crop Pest Commission published the results of an experiment that offered great promise to farmers in search of an effective weevil poison. Newell found that powdered lead arsenate killed the beetles without damaging the cotton. Word of the experiment spread quickly around the South. In the Delta, the Greenville *Daily Times* hastily reported that "recent experiments with a new kind of poison" indicate that boll weevil "control may be obtained." Fox and Ewing read Newell's reports and shared the scientist's optimism, but they also recognized that Paris green's major shortcoming wasn't its damage to cotton, but its application. This was still an issue with powdered lead arsenate. In Newell's successful experiments, workers had spread the poison by hand—forcing it into each infected square with a "dust gun." It was impractical from a labor and cost standpoint to think that DPLC (or any other farm operation, small or large) would be able to apply the poison efficiently or effectively in that manner.[31]

Buoyed by the poison's effectiveness, DPLC's scientists set after a solution to the application problem. In 1915, Ewing rode through the region interviewing farmers who had begun using lead arsenate. He visited Bert Coad, a USDA researcher based across the Mississippi River in Tallulah, Louisiana, who had performed successful experiments with lead arsenate on a large scale. Seeing Coad's success, Ewing became convinced that with DPLC's size and resources, it could afford to research a solution to the application problem. In the spring of 1916, the plantation performed its first field tests on lead arsenate, as well as a similar compound, calcium arsenate, which had proven even more effective against the weevil. In 1917, Coad dispensed two of his federal researchers to DPLC so that the company could direct more extensive tests over even larger areas. Coad recognized that the government's research could benefit from DPLC's size and labor force; DPLC, on the other hand, was happy to have what amounted to federally funded research conducted on its own land.[32]

Coad and his assistants began developing simple mechanized solutions for the application problem. During the first experimental season, laborers applied pesticides by hand. A tenant would walk down a row and spray each plant individually with the poison. As with Coad's experiments in Louisiana, this method proved effective on experimental plots, but was far too time consuming and labor intensive to be put to use on all of DPLC's thousands of acres. Plus, the spray guns constantly broke. For the 1917 tests, men rode mules up and down the rows, but still sprayed each plant one-by-one as they went. Even with these advances, farmers still had problems getting the poison to stick to the plants. As a result, they began spraying at night; the dew acted as a glue to adhere the powder to the cotton. "We done it at night," one manager recalled, "sometime you'd go until about eight o'clock in the morning."[33]

In 1918, Ewing came upon a new idea: a dusting machine mounted to a two-wheel cart pulled by mules. Ernest Haywood, one of DPLC's early managers, described how it worked: "We had a little old motor up there we would crank it and we had a hopper and we'd put dust in the hopper. We . . . set the gage and . . . hitch a mule and you run them things all over the place and then we had carbide lights and you charge that thing up." These machines were powered by the traction of the cart as it moved through the field. A fan would blow the dust from the platform onto the plants as it passed. But this solution too had its problems. The poison usually settled onto the ground rather than the plants, and if there was any breeze at all, the poison was impossible to direct. By 1920, DPLC was poisoning all of its cotton by machine, but problems persisted. Despite its ability to pay for the poison, the plantation's fight to stop weevil damage with insecticide had proven wasteful and inefficient.[34]

The final solution to the application problem came from above. Shortly after World War I, the USDA began funding experiments on aerial application of poisons at the Delta Research Station in Tallulah. Bert Coad, the station's chief scientist, found that by toting finely powdered calcium arsenate in a hopper underneath an airplane, pilots could swoop down over a cotton field, flying sometimes fewer than a foot above the crop, pull a lever releasing the poison, and spread the insecticide onto the cotton. The force of the plane's movement blew the dust onto the plant with greater strength than could be achieved at the field level. The Army Air Service eventually contributed surplus World War I planes to the cause and the first experiments in air crop dusting cotton were born.[35]

Not surprisingly, the first private farm on which Coad conducted aerial crop dusting experiments was DPLC. Coad recognized that the company's long, uninterrupted fields would be a perfect place to experiment with aerial pesticide application. "Everybody laughed at the idea [of crop dusting by airplane] and scorned it at first," Coad recalled, "but our old friends at Scott still stood behind me." DPLC recognized that if Coad's scheme were possible, it would make poisoning weevils more effective and efficient. In the early 1920s there were still risks and high costs associated with crop dusting, but the promise of a quick and effective means of poisoning boll weevils was too attractive for DPLC not to explore.[36]

In 1926, Coad led the first public demonstration of boll weevil dusting by plane at DPLC. The prospect of using airplanes to poison weevils was odd enough to attract the attention of farmers from across the Delta. "All through the night people were arriving in Scott from as far away as Memphis" to witness the morning demonstration, Coad remembered. The pilots, sent over from the Army Air Service station, had been "a little too well entertained the previous evening" and looked to Coad to be "very groggy" when they climbed into the dusters. With the crowds amassed waiting to witness the spectacle of crop dusting, however, nothing was going to derail the performance, no matter the pilots' hangovers. Coad loaded the planes' hoppers with calcium arsenate, and soon they were skimming across DPLC's fields followed by a cloud a white powder. The demonstration went off without a hitch. Fox and Ewing were so impressed with the early experiments that they pledged the company's continued financial support, and promised to hire any company that would offer effective crop dusting services.[37]

It did not take long for a fledging dusting operation to answer DPLC's offer. Collett E. Woolman, an agent with the Louisiana extension service, had observed Coad's experiments since the early 1920s. Woolman, not unlike Fox and Ewing, decided to leave his position with the government sector to help a private company sell a strategy for beating the boll weevil. His knowledge of farming, combined with an infectious personality, made him the ideal spokesman for crop dusting. He joined the Huff Daland Dusters Company and moved its base of operations from Georgia to Monroe, Louisiana, to be nearer Coad's experiments. In 1928, the company had such success in aerial boll weevil control that the dusting division separated from its parent company to become Delta Air Service. The company began dusting cotton throughout the region, but its main client was DPLC. By the early 1930s, Delta Air billed DPLC as much as $11,000 per month for its dusting services. Even as the pesticide of choice changed from the 1920s on, airplanes still proved to be the most effective means for accurately spreading poison on cotton. (Eventually, Delta Air developed a passenger service in addition to its dusting operation and became one of the biggest airlines in the world.)[38]

Whether poisoning for the boll weevil with a handheld spray gun, a machine in the field, or an airplane, pesticides fell on more than just the cotton plant. Though the barnstorming duster pilots were known for their ability to skirt the tree lines in order to spray poison into the farthest corners of a field, they usually applied those skills to guaranteeing full coverage, not to avoiding the buildings that dotted the landscape. In fact, DPLC was a prime experimental plot because of its long, wide fields that allowed planes to spread poison in extended rows without frequent turns. The sharecropping cabins located in the middle of the fields were of no concern to the pilots, nor to DPLC's managers. Bert Coad admitted that "cabins frequently were subjected to a cloud of dust," but, he argued, "the poison . . . is so thin . . . that the portion drifting to any other point does not settle in injurious quantities."[39]

The use of sharecroppers in spreading boll weevil poison and the disregard for their safety during aerial spray runs exemplifies the overall treatment that laborers received at DPLC while the company fought the boll weevil. One manager, Dick Holman, told an interviewer that spreading the arsenical compounds could be dangerous for wildlife and the boll weevil, but refused to admit that it had an effect on humans. "You used to kill all the cows and deer and everything else," with the poison, Holman argued, but sharecroppers "would stay out of the way of it." Unless, of course, they ate it—a story many landowners commonly told as a way of explaining that the poison was safe. When applying the pesticides, Holman admitted, "them niggers run across a watermelon patch, he would stop there and eat them watermelons just like he was eating at the house. He never did pay no attention to that poison. And I've never heard of anyone getting hurt yet." Others told stories of farmers mistaking the powdered poison for flour and baking it as bread. Early Ewing Jr. assured an interviewer in 1974 that calcium arsenate "wasn't too toxic, unless you made biscuits out of it, which some wives did, on a few occasions. And so it wasn't too toxic to the people that use it, like modern-day insecticides are."[40]

Today's scientists disagree. Experts now recognize both the dangers of immediate exposure to calcium arsenate and the long-term effects, but the discovery of risks to humans was not a twenty-first-century revelation. As early as the 1890s there was great debate in the international medical community over the health risks of arsenic. Today, calcium arsenate, an inorganic chemical compound, is classified as a medium to "very high risk" poison for oral exposure and medium risk for dermal contact; unprotected long-term exposure to this compound produces a surfeit of health problems. DPLC's workers had no protection from the poison. In fact, after a night spent dusting by mule, workers went home covered in the dust from head to toe. Subjected to these chemicals day and night, tenants experienced a range of immediate ailments, including gastrointestinal pain, diarrhea, irritability, headache, drowsiness, confusion, vomiting, and toxic psychosis. The effects of long-term exposure, which include a range of diseases from dermatological conditions to a host of cancers, have gone unrecorded. And though DPLC made much of the presence of a company doctor in Scott, the relative powerlessness of black sharecroppers over the medical treatment they received and the presumed unwillingness of DPLC doctors to point to poisons as the cause of worker's sickness conspired to make the doctor's presence all but meaningless. If sick, tenants had to first approach the unit managers to get permission to make an appointment with the company physician. Managers regularly denied these requests. In addition, trips to the company doctor were not free; tenants signed over cotton slips to the physician to pay for his service, promising a portion of the cropper's cotton at the end of the season. Many tenants simply refused to pay to see the doctor no matter how bad they felt.[41]

Perhaps the most telling aspect of DPLC's boll weevil poisoning program was that the bulk of the costs of the insecticides were actually passed on to sharecroppers themselves. DPLC bought enormous quantities of the poisons at wholesale costs, then charged each tenant half the cost of the insecticide based on the cropper's acreage. The company of course did not allow tenants to forgo buying the poison, and before the evolution of aerial application, the responsibility of paying for and applying it fell on the croppers. Even once the company was paying Delta Air to dust the fields, the cost was passed onto the tenants based on their individual acreage. This policy is indicative of DPLC's position as both a modern corporation employing the latest advances that science and technology had to offer, on one hand, and, on the other, its having to operate with the cooperation of a vast human workforce.[42]

*

For all the new technologies that DPLC's size and capital helped to develop, the mega-plantation could not reinvent the basic manner of seed planting, weed chopping, and cotton picking—in other words, the work of making cotton. At least until the late 1930s these tasks depended on people, not machines. For the first thirty-odd years of the twentieth century, despite DPLC's focus on science and mechanization, the single factor on which cotton cultivation most depended was human labor. Like many large plantations in the South, DPLC had a foot in two separate worlds, one modern and one traditional.

The company's dependence on the overwhelmingly black workforce frustrated its managers. Jesse Fox, DPLC's superintendent, summed up his feelings about labor when he told his managers his basic theory of cotton planting: "Well gentlemen, I just want to tell, it takes niggers and mules to make cotton." Dick Holman, a longtime manager of DPLC, echoed this recollection. Holman reduced DPLC's size and resources to a simple observation that the plantation's success rested on "a good many mules and . . . worlds of niggers."[43]

Despite their belief in the Delta's limitless environmental bounty and inexhaustible supply of labor, the boll weevil threatened Fox and Holman's understanding of cotton farming. When the beetle first appeared at DPLC, and then continued to wake from hibernation spring after spring, seed programs and pesticides became relatively more important to the company's bottom line than producing a crop. But none of these developments lessened DPLC's dependence on human labor. In fact, the development of new farming techniques forced DPLC managers to pay even more attention to the way tenants planted, fertilized, applied poison, and picked the crop.

Fox, Ewing, and the other managers had developed a rather complicated method for farming and cultivation, which necessitated a system where laborers would carry out their plans to the letter. This had evolved into an amended system of sharecropping where managers kept watch over tenant activity, attempting to strictly control the manner in which they farmed. The boll weevil made the details of the day-to-day work in the fields more important to the managers. Fox dictated how the soil was to be prepared, the schedule of planting, the spacing of plants, and the timing of chopping, poisoning, and picking. From the company's point of view, sharecropping not only gave managers strict control over tenants' work, but it allowed tenants a share of the risk and cost. Even in seasons when weather or the boll weevil shortened production, the company hedged its stake with tenants' own investments. Croppers provided labor, but also consumed goods in DPLC stores.

As a body of historical work suggests, sharecropping on large plantations was a system that produced profits for the landowners in a variety of ways. One study of Delta farming conducted in 1916 found that sharecropping all but guaranteed Delta planters a return on their labor investment of between 6 and 18 percent. From DPLC's perspective, farming in the presence of the boll weevil could not only be a manageable endeavor, but a profitable one. DPLC padded their investment in labor and cotton with their sales of seed and the profits generated by tenants' accounts in the company stores.[44]

From the tenants' perspective, working for the country's largest cotton plantation had some advantages over smaller operations, especially considering the appearance of the boll weevil. Hundreds of sharecropping families had moved north from the infested territories into the Delta about the time DPLC was building its workforce. The company also sent recruiters into Texas, Louisiana, and parts of Mississippi where the weevil had done the most damage. Lillie Belle Parker recalled that in 1914 her parents, then living in Magee, Mississippi, met "a man going around getting labor for the Delta Pine Land." "And, he told my daddy that they was giving them a house, residence, furniture in the house and giving them a mule to work the place and that's why he was up here," Parker told an interviewer.

For sharecroppers like the Parker family, life at DPLC offered some hope. The company's size suggested to many tenants that it would be a stable place to live and work. The promises of housing, medical attention, and community fostered the dream that they might work for a few years and make enough money to buy their own land. As a result, thousands of croppers moved to Scott in search of work; tens of thousands of laborers would eventually call DPLC home. Early Ewing told his son, "every Negro and white man in the Delta at one time or another had worked for Delta and Pine Land Company."[45]

The hope that DPLC, and the Delta as a whole, offered black Mississippians in search of work and community was fleeting. Though industrial planters adapted to the boll weevil, passing on much of the increased work created by the pest to the tenants, and in some cases even profiting from the pest's spread, sharecroppers usually felt only increased pressure and decreased hope as the weevil appeared. The beetle slowly shut the door on the dreams of Delta tenants, as DPLC ratcheted tighter its management and supervision of its employees' work and social lives.

When they arrived, sharecroppers found work at DPLC strictly controlled. Sharecroppers on DPLC's sixteen sub-plantations woke six days a week before daylight to the ringing of the company bell. Each of the firm's sixteen farms was further divided into units, each having its own set of barns housing mules, tools, feed, seed, and other supplies. After breakfast, prepared mostly by women and eaten in workers' company-owned cabins, croppers walked to the unit barn to pick up their mules and, depending on the time of the season, walked out into their plots and began cultivating, planting, chopping, or picking. Women and children worked in the fields at various points in the season, but their labor was most important during chopping (late spring) and picking time (late summer and early fall). On average, each family worked ten acres of cotton for each male head of household and an additional five acres for each able-bodied person in the family.[46]

At its height, in the early 1920s, there were 1,400 sharecropper families living and working at DPLC, though the conditions of their employment—and labor's effect on DPLC's bottom line—were never static. Tenants had accounts at the company store, where they charged food, clothing, farming supplies, toys, and anything else that DPLC decided to provide. The company charged interest on each purchase and tenants would settle their accounts at the end of the season. Despite the common assumption that sharecroppers always came out further in debt at the end of the season than when they began, most plantation records suggest that only in dire times did the majority of tenants end a year in debt.[47]

Depending on several factors, including most importantly the size of the tenant's harvest and the price he or she could get for their cotton, tenants finished most seasons ahead. In years of heavy boll weevil infestation, however, tenants bore the brunt of the costs for pesticides, fertilizers, and additional labor, and stood to suffer the most if the crop was diminished. In years when the insects destroyed the crop despite tenants' application of poison, the croppers still paid for the poison with the scant cotton that the boll weevil left. As a result, in years when the boll weevil population was high, DPLC

was assured of increased action at the company store. Of course, the plantation had to bear the wholesale cost of fertilizer and pesticide and hope that
farmers made a sufficient crop at the end of the season to pay back what they
had used, but in years when tenants were successful, DPLC not only realized profit on the cotton itself, but on the supplies that they forced tenants
to use in the first place. Fledgling business magazine *Fortune*, which usually
focused on northern industry, even remarked on DPLC's capacity to extract
every cent of profit from the land and its workers. The magazine applauded
then company president Oscar Johnston's "maximum expertness in cotton
management," and called the firm's gross earnings "freakish."[48]

DPLC's strict administration of workers' lives, the company's control
of where tenants lived, and how and when they farmed, was guided by its
exacting management style, a style girded by racism. In one sense, the record of racism left by the company's managers was typical of white Mississippians in the era of Jim Crow, but in the context of the company's efforts
to build a model of industrial agriculture, the comments take on a class and
labor component. "A nigger ain't got no class," claimed one former DPLC
manager years after his retirement. "They don't care how you dress them
up. No matter how you dress them up he's still going to be a nigger." For
the white workers at DPLC, blackness and sharecropping were one and the
same and the industrial structure of the farm gave that equation different
outcomes from its farm neighbors. In fact, the paternalism most associated
with Delta plantations—an attitude rooted in planters' fabricated idea that
they were looking after sharecroppers like their own children—was absent at
DPLC.[49]

If the racism was ratcheted up at this industrial farm, so too was labor's
daily resistance to that racism. Though the paternalism that characterized
most large Delta plantations was certainly little better for black sharecroppers, DPLC's brand of industrial management created a kind of racial supremacy with fewer of the personal relationships between laborers and owners that developed on smaller plantations. Despite the white supremacy at the
heart of Delta society, or perhaps in conjunction with it, white landowners
and managers were forced to rely on black labor, and this dependence gave
the workers a modicum of control over their conditions. Landowners and
managers simply could not restrict every aspect of black farm life. As outlined
in earlier chapters, planters' obsession with labor is evidence in itself of the
limited power workers had over their bosses. In short, planters needed to
obsess about labor; they threw tenants parties, extorted and lynched them
because workers needed to be controlled. Without these constraints (and to

a degree even with the planters' oppression in effect), tenants were active and constantly moving. Like the boll weevil, workers were in constant motion.

As in Texas and Louisiana, the social lives of Delta cotton workers directly related to the ever-changing status of their field work. Perhaps the most important way workers resisted the strictures of farm life was through participation in a complex social life. Sharecroppers went out, socialized, and spent money to exercise a modicum of control over the drudgery of their work lives. How tenants spent money, worshipped, traveled, sang, and let off steam — particularly in the first two decades of the twentieth century — is a direct reflection of the boll weevil's effect on Delta cotton culture. The blues, especially the dozens of songs directly and tangentially about the boll weevil itself, serves as a historical record of the boll weevil's invasion of the Delta and proof of the region's unfulfilled promise. This musical form is inseparable from the work and social lives of black tenants in the Delta and explains something about how sharecroppers felt about the pest, about landowners and about their hope for a better life.

Most weekends during the cotton season, sharecroppers from DPLC rode in cars or wagons to town on Saturday night to see and be seen, and to enjoy some entertainment. Ewing recalled that workers often had a little cash at the end of the week from "non-cotton related work they might have done," and they ventured into town to be entertained. "They had tent picture shows and things like that," Ewing remembered. "I know they used to have . . . a minstrel show. Brer Rabbit Minstrel Show and Silas Green used to come through here . . . when people had a little money."[50] Most of the entertainment enjoyed by sharecroppers was not of the tent show variety, however. Most weekends tenants did not need to venture into towns to find entertainment.

At the center of plantation social life was the bluesman. Far from the mythical travelling singers of American folklore who wandered through the South making deals with the devil, most Delta bluesmen were professionals in search of selling their entertainment to the black community. They did so in a space on or near a plantation, all but closed off to white farm managers, where farmers could gather to hear the blues and enjoy time off. In this environment bluesmen became an important part of the entire community's social life. Opportunities to perform for pay were common enough that talented musicians could make a living outside of the cotton fields by moving from one party to the next, though most never strayed too far from the rural cotton communities that formed the base of their audience. "Every singer I've ever ran across, he is not a worker," recalled one manager of blues acts. Nevertheless, the bluesmen became *representations* of cotton laborers themselves. As James C. Cobb argues, a traveling musician was "a key figure, symbolic of a communal culture . . . [who] entertained his audiences by expressing deeply felt, shared emotions in a manner that made him more than an entertainer."[51]

Although many blues songs expressed universal themes of love and loss, those about the boll weevil were anchored in the specifics of that time and place. From the historians' standpoint, the boll weevil songs that arose in this space are complex documents of life with the pest. To be sure, they were sharecroppers' reflections on what the insect was doing and how it had affected their experience. But the songs document reception as well. Audiences applauded, nodded, and shouted in agreement when a song struck a familiar chord, as the boll weevil songs did. As a whole, the boll weevil blues that emerged from the early-twentieth-century Delta also speak to the power of the boll weevil myth. The tunes arrived there before the pest itself did, communicating information about the bug and the changes it promised to bring. In turn, the songs propagated these understandings of the weevil; they spread word of the pest around the South (and eventually around the world).

*

Perhaps no single bluesman in the Delta got more people yelling and more hips shaking than Charley Patton, and it was his own vision of the boll weevil's southern journey that catapulted the singer to the national and international stage. From a young age, Patton found work in Mississippi's fields a bore. He tried to play his guitar more and work in the cotton less. His family moved around Mississippi during his childhood, but around the turn of the century, the Pattons moved north to the Delta, and settled on the Dockery plantation just east of Cleveland, about thirty-five miles from DPLC. Bill Patton, Charley's father, sharecropped for three years until he had saved enough money to buy land near Renova, an all-African-American community where he eventually built and ran a general store. Charley stayed behind in Dockery, where he began life as a bluesman.[52]

Patton had probably begun playing music prior to his arrival in the Delta, but his musical interest flourished at Dockery. He played with and learned from other tenants like Toby Bonds and Henry Sloan. He also worked some in the fields, though he tried to avoid it. Though the details of his career are sketchy, Patton most likely began playing professionally around 1907, just as fear of the encroaching boll weevil reached a fevered pitch.[53]

By 1928, Patton had earned the reputation as the best and most active blues singer in the Delta. He played at parties throughout the cotton territory for whites as well as blacks. Perhaps he felt he had reached the peak of his local fame when he boarded a train in 1928 for Jackson. He had heard from another musician that there was a white man there, H. C. Speir, a squat, cigar-smoking drug store owner who recorded singers in a makeshift studio above his pharmacy. Patton found Speir, played a few songs for him and was soon standing in front of a microphone in the upstairs studio recording his first tracks. Speir must have like what he heard because a few months later he drove up Highway 61 into the Delta with a primitive recording contraption in his car trunk, looking to record Patton again. He found the bluesman in a cabin and made more records. This session would change the rural black Mississippian's life forever. Only one month later, Patton was hundreds of miles from Dockery, laying down his first track for Paramount Records: "Mississippi Bo Weavil Blues."[54]

The recording, released by Paramount later that year for nationwide distribution, was the first widely disseminated commercial recording of a boll weevil song. Its effect was great. The record found an audience throughout the South, but it also made its way into jukeboxes and record collections across the country. People who had already experienced the boll weevil's damage firsthand identified with the singer's plight, while those who lived far from cotton fields learned of the seriousness of the tiny insect's spread. But more than simply a cultural artifact of the human-pest interaction, Patton's "Mississippi Bo Weavil Blues" is an entrée into the world of sharecropping in the Delta. It is a song about mobility, cultural space, economic opportunity in the face of the insect threat, and white peoples' commitment to a race-based labor system wedded to the constricting arrangements of tenant farming.

Those who played the record on home phonographs, listened in a juke joint or club, or live from Patton himself, heard a halting, cryptic song that for many must have been confusing. To a listener today, it can be almost unintelligible. First, one must spend a lot of time with the song to understand the words. The recording is rough and crackly and Patton's voice pierces through the static with a low-pitched throaty roar. Understanding his voice is not all the fault of the primitive recording; Patton was never known for his vocal clarity. In the 1920s, bluesman Son House heard Patton play at the Dockery plantation several times and remembered that "a lot of Charley's words . . . you can be sitting right under him [and] you can't hardly understand him." Booker Miller, another blues singer, remembered that Patton had a "growl in his voice" that made it hard to understand.[55]

Once a listener becomes accustomed to Patton's voice, the story he tells is still confusing. At some points in "Mississippi Bo Weavil Blues" he sends clues as to who is speaking, with lines that begin "farmer said" or "boll weevil said," but in many of the stanzas he switches from the voice of one character to another without warning. Even an ear acclimated to his voice and the crackly recording cannot always be sure who is speaking to whom. Over a single-chord guitar line that provides at once a low rhythm and a high, almost squealing jingle, Patton begins singing, "There's a little boll weevil, see him movin' in the air, Lordy/You can plant your cotton and you won't get a half-a-cent, Lordy." Listeners across the country who tuned in on the radio, picked up the record in a store, or caught Patton live, heard a voice, in the words of novelist Tom Piazza, "flaring up and dying down like a kitchen match."[56]

Without altering the phrasing or tone of his giant voice, Patton creates two characters, a farmer and a boll weevil, having a conversation. "Boll weevil, boll weevil, where's your native home?" the farmer asks. "A-Louisiana r-an' Texas is a-where I's bred an' born," the pest responds. But the boll weevil was on the move. "Boll weevil lef' Texas," the farmer explains to the listener, "he biddin' me farewell." Then a plainly spoken question for the pest: "Where you goin' now?" The weevil boasts "I'm going down to Mississippi gonna give Louisiana hell, Lordy!" Patton's role then switches to narrator, explaining that the "boll weevil told his wife, I believe I may go north." The threat, or promise, of a move north is too much for Patton's narrator to bear. After a few pointed plucks of a high note on his guitar, he delivers a slow, un-sung, haunting promise: "I won't tell nobody."[57]

This secretive promise hints that there is more going on in this song than a boll weevil eating some cotton. The confusion over who is speaking, the constant shift of the narrator, is intentional. This deliberate misdirection blurs not only the line between farmer and pest, but between singer and subject, between hero and villain. The song can take on a different meaning, then, if we think of the pest character not only as the boll weevil but also as the landless tenant farmer. If it is a sharecropper telling his wife "I believe I may go north," then the secrecy that Patton promises with his line "I won't tell nobody" takes on a whole new gravity. Patton's character's threat to move, or even simply his allusion to the boll weevil's own frenetic migration, was at the heart of tenant farmers' lives. The speaker's whispered threat to move north spoke to a long history of movement of laborers into, within, and from the Delta as well as to an oppressive and violent tradition of landowners attempting to restrict that movement. Though Patton's version does not contain the line "just lookin' for a home," which was common to the majority of published and recorded songs about the pest, there is no doubt that the insect and farmer he describes are searching for a better life somewhere else.[58]

Migration is an important theme in Patton and his contemporaries' boll weevil blues, and although the common narrative of the Great Migration argues that World War I was the catalyst for a long, steady exodus of black southerners to northern cities, the picture that emerges from the Delta in the first decades of the twentieth century suggests a more complicated story. Black migration within and from rural Mississippi was anything but a linear phenomenon. World War I created enticing jobs in the North, and the boll weevil certainly made life more difficult for tenant farmers, but there was seldom a single reason for people leaving the rural South; there were as many reasons for leaving as the number of people who left. In the interwar years, for instance, the exodus from the Delta did not even mean a net loss of African American population in the region. As J. William Harris points out, black laborers left northwestern Mississippi in substantial numbers from 1920 to 1925, but the overall rural black population did not drop at an equal rate. From 1926 through 1930, the total number of black sharecroppers in the Delta *increased* to its highest level ever: 77,000 tenants. The promise of the Delta that had attracted landless laborers since the turn of the century had not dissolved by the First World War. In fact, in the 1920s, farm conditions in much of the South lying east of the Mississippi River were deteriorating, but the Delta remained an attractive place for resettlement. In the 1910s and 1920s the weevil spread east to the older lands of the Cotton Belt and laborers there looked west, as well as north, for places to resettle. Tenants in Virginia, South Carolina, and Georgia saw Mississippi as an attractive place to settle until the onset of the Great Depression.[59]

Part of the reason for historians' incomplete picture of this migration is the mythological power of the boll weevil. The idea that farmers were powerless against the pest and that it destroyed everything in its path contributed to the notion that when the insect arrived sharecroppers had no work to do, saw certain poverty, and fled. Not only have historians miscalculated the relationship between the boll weevil and migration, the historical record left by those who did not migrate often gives the impression that the boll weevil was the main factor for migrants.[60]

White Mississippians, for instance tended to cite the pest as the reason for tenant emigration. Many chose to point to this nonhuman factor as the source of worker discontent rather than those aspects of the tenant life that the landowners themselves had created. In fact, white farmers seemed genuinely shocked as laborers left. They used violence and peonage to fight that migration, and continued to argue that they had been *abandoned* by "their" workforce long after the migration slowed. As one Greenville businessman recalled, some scholars believed the adoption of mechanized farm equipment in the 1930s and 1940s had forced black labor from the fields. He argued instead that the machines replaced, rather than displaced, workers. By the time tractors arrived, black workers were "already gone . . . And you couldn't stop them from leaving." Which is not to say that whites did not try to stop to stem the tide.[61]

*

The legacy of the boll weevil in the Mississippi Delta goes beyond the themes of migration and mechanization. As the songs about the boll weevil and the memories of black farmers and white landowners reveal, the insect fundamentally changed the way that farmers—both landowners and tenants—thought about their work, but it didn't upset the basic system of cotton farming. Historian William Lincoln Giles correctly points out that in the late nineteenth century, "Large farming operations differed from small ones only in having more hands, more hoes, and more mules."[62] It took until the 1930s for that to no longer be true. The weevil did make cotton farming more difficult for everyone—as Figure 9 shows, the pest destroyed significant amounts of

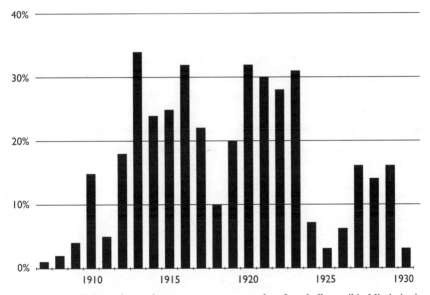

FIGURE 9. Estimated annual percentage cotton crop loss from boll weevil in Mississippi, 1907–1930. Illustration by the author. Data from Willard A. Dickerson et al., eds., *Boll Weevil Eradication in the United States through 1999*, the Cotton Foundation Reference Book Series, no. 6 (Memphis: Cotton Foundation Publisher, 2001), 614–15.

cotton statewide—but the underlying structure of labor, credit, and land use that made cotton king of the Delta remained in place.

The boll weevil did change aspects of life in the Delta, but it did so indirectly. The pest aided the development of Delta industrial plantations like DPLC. These companies used fear of the weevil to sell new ways of fighting the pest, thus guaranteeing their own profitability through marketing weevil-beating seeds to the rest of the South. The company's success did not happen overnight, but the weevil's long-term presence drove its continued experimentation. By the mid-1930s DPLC could boast that it sold more quick-maturing weevil-beating cottonseed than any other firm in the world.[63]

Despite the growth of industrial plantations, twenty years after the weevil's arrival, white landowners still ruled economically and politically and the cotton workforce was still overwhelmingly black and poor. There was no major outmigration of workers because of the weevil; the tenant population actually increased. In fact, the pest disrupted planter hegemony only briefly. That the croppers on Johanna Reiser's plantation became renters because of the pest was by far the exception; in general, the arrival of the boll weevil meant increased landowner control and scrutiny of workers' personal and work lives.

It is telling, however, that the single aspect of this period's history that has had the most lasting relevance to southern history is the blues culture born in the Delta when the boll weevil first appeared. Through this social atmosphere and the songs and stories that it created, an alternative picture of the power of the boll weevil emerges. Not only do the songs record the obstacles to the day-to-day lives for Mississippi's black sharecroppers; they also document the ways in which ideas about the boll weevil were inherited, digested, and retold for audiences across the country. Hearing of the pest in song was an important way that farmers—black and white—who lived to the east of the pest's advancing front knew what to expect from the beetle.

CHAPTER 5

"The Herald of Prosperity": The Promise of Diversification in Alabama

In January 1910, as the first boll weevils crossed the Mississippi border into Alabama's southwestern corner, the *New York Times* published a lengthy article about the pest's trek and southerners' unsuccessful efforts to stop it. Reiterating the by then well-known and untrue story of the weevil's all-encompassing destructiveness, the *Times* claimed there was simply no way to beat the bug. The weevil was "an illustration of the immutable forces of nature," according to the paper, "and the futility of man's feeble efforts against a force of the sort." There were no poisons that could stop the pest, no effective natural enemies. "Where the weevil comes, he stays," the *Times* maintained, "what the weevil gets he holds."[1]

Articles like these not only perpetuated the myth of the weevil's sweeping, comprehensive devastation, but also concurrently endorsed the insect as a harbinger of an agricultural revolution. The spread of the boll weevil could mean that crop diversification would finally take hold in the South. The only way to stop the bug, the *Times* explained, was to take away its sole source of food. Nothing short of pulling up the region's cotton plants and planting something else—anything else—could limit the pest's damage and improve the lot of southern farmers big and small. Would-be reformers and landowners had talked of diversification for generations, but each spring they watched as farmers devoted more and more acreage to the fleecy white crop. The paper maintained that despite southern landowners' widespread, long-term commitment to cotton, their sentiment against planting beans or vegetables and raising stock was actually waning. "The Deadly Boll Weevil" was sweeping through the South, the headline beamed, "Bringing Revolution With Him." Some farmers were even calling it the "Prosperity Bug." The *Times* suggested that the tiny insect was accomplishing what generations of modern voices could not.[2] The newspaper's lengthy article presented the reader with a hypothetical southern landowner, who, within the weevil's sights, would cut his cotton acreage by a third, plant more oats in the spring, and more corn, cow peas, and pumpkins in the summer. "He gets hogs and cattle to raise and breed," the paper imagined, producing on his own land "chickens and eggs and meat and meal and molasses." The farmer would never need to rely on merchants or banks again, growing cotton only on excess land. "What cotton he gathers after his wrestle with the boll weevil is clean, clear profit." No longer buying his meat from Chicago, flour from Minneapolis, and nails from Pittsburgh, "he is as nearly an absolutely independent man as it is possible to be." The message of the parable was clear: the boll weevil would teach southern farmers once and for all that dependence on cotton was suicide. Diversification would end rural poverty across the South.[3]

But this New York dream had little grounding in Alabama reality. From the point when the boll weevil crossed the Mexican border in the late nineteenth century to the publication of the *Times* article in 1910, it had traveled more than eight hundred miles and destroyed an estimated 1.6 million bales of cotton, worth $107 million. Yet, in spite of these figures, southern farmers were not abandoning the plant en masse. As the previous chapters have shown, in parts of the South, farmers were actually *expanding* the amount of land devoted to the fiber. Despite the substantive damage caused by the weevil from 1892 to 1910, Texas had doubled its cotton acreage.[4] Though Louisiana and Mississippi saw modest drops in cotton acreage statewide after the weevil's initial invasion, areas within these states increased their commitment to the staple. Planters in the Delta increased both cotton acreage and cotton yield during the first decade of the weevil's presence. Farmers were actually planting more cotton and getting more lint from each plant than ever before. In 1910, the pest stood on the Alabama border, as farmers, businesspeople, and government experts wondered what changes the pest would exact in their state.[5]

From the perspective of the boll weevil, Alabama bore little resemblance to the Delta. Whereas the dark topsoil of the alluvial Mississippi region is rich and flat and consistent, Alabama has a landscape and geology that can vary and shift from acre to acre and field to field.[6] The southeastern corner of the state is worthy of particular study because of the fascinating combination of historical forces at work as the boll weevil approached. (See Figure 10.) The

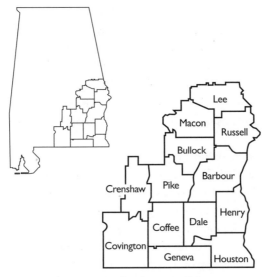

FIGURE 10. Southeastern Alabama.

region's geography embodies two important southern cotton-growing areas, the Black Belt and Wiregrass. Southeastern Alabama was also home to two of the South's most important agricultural schools, Tuskegee Institute and Alabama Polytechnic Institute (renamed Auburn University in 1960).

In addition to the fundamental differences in the region's agricultural environment, the timing of the insect's arrival in Alabama was important. The women and men there were not surprised by the boll weevil—they had heard about, read of, and researched the pest for fifteen years before it arrived in their state. Alabamans had time to prepare for the weevil's arrival. The combination of these two factors meant that, unlike in any other region where the boll weevil had appeared to that point, diversification had a genuine chance of success in Alabama. Knowing the range of effects that the boll weevil might have on their society, southeast Alabamans were actually interested in the idea of growing something other than cotton. It was a region wooed by the promise of diversification in the face of the weevil's menace, and one in which an amazing and unlikely combination of factors made the dream of diversification come true.[7]

*

As early as the 1820s, farmers in the southeast corner of Alabama tried to make cotton grow in both the Black Belt and Wiregrass with varying results.

In the naturally fertile and relatively flat Black Belt counties, cotton grew with little effort. In the Wiregrass, however, the soil was too sandy and lacked sufficient nutrients to produce the plant. In the second half of the nineteenth century, however, the advent of widely available commercial fertilizers made cotton farming in the Wiregrass possible. By the turn of the century, farmers in southeast Alabama were committed to cotton production everywhere and anywhere the staple would grow, no matter the soil type or the impact it had on the land.[8]

Immigrants intent on turning the poor land into a cotton kingdom flooded southeastern Alabama. From 1890 to 1900 Covington County, a sprawling district on the Florida border, saw its population more than double; Geneva and Coffee Counties experienced dramatic population gains as well. In fact, all counties in the region except one experienced double-digit increases in population during this first decade of cotton's viability.[9]

Despite the availability of fertilizer and local farmers' willingness to pay high prices for it, cotton farming in the Wiregrass was never a sure thing. Farmers there were more dependent on sufficient rainfall, field preparation, and precise application of the proper fertilizer than were growers in the Black Belt. A common Wiregrass rhyme sums up the precariousness of their plight: "Sift your meal and save your bran, 'cause you can't make a livin' in sandy land."[10]

As happened across the South when marginal lands caught cotton fever, most of the immigrants to southeast Alabama soon found themselves mired in tenancy. As Table 1 shows, in 1910, on the eve of the weevil's entry into the state, those southeastern Alabama counties that grew the most cotton also had the highest percentages of tenancy, as well as the largest African American population. Bullock County, for instance, located in the Black Belt just to the east of Montgomery, had a tenancy rate pushing 90 percent; not surprisingly, given that figure, 80 percent of its population was African American and it was the region's most productive cotton land. Of the six counties with tenancy rates higher than the state average, all were major cotton producers. Race, tenancy, and soil combined to determine the response of each county to the boll weevil's arrival.[11]

In 1910, southeastern Alabama was home to more than expanding cotton production and growing rates of tenancy. Alabama Polytechnic Institute and Tuskegee Institute, the state's two leading agricultural colleges, had been in existence long before the boll weevil's arrival in the region, but each would grow significantly after (and because of) the pest's appearance in the state. From 1910 to 1930, Alabama Polytechnic and Tuskegee developed into two

TABLE 1. Cotton, tenancy, and race in southeastern Alabama, 1910.

County	Cotton acreage	Percentage of farms operated by tenants	African Americans as percentage of total population
Bullock	107,099	87.2	81.70
Barbour	99,170	70.3	63.64
Pike	96,540	64.1	42.76
Macon	89,796	83.3	81.61
Russell	83,750	80.1	78.10
Lee	79,261	71.0	59.91
Coffee	72,535	45.2	20.18
Henry	70,229	50.2	37.64
Dale	61,056	45.2	22.98
Crenshaw	58,833	47.3	28.45
Geneva	56,645	35.5	16.85
Covington	42,528	21.3	15.86
Total for state	3,730,482	57.7	45.25

Source: *Thirteenth Census of the United States Taken in the Year 1910, vol. 1: Population; vol. 6: Agriculture, 1909 and 1910* (Washington, D.C.: Government Printing Office, 1913).

of the South's most influential agricultural schools and became major voices for crop diversification.

In 1872, with a charter from the state legislature and federal Morrill Act funding, Alabama Polytechnic took over operation of the small East Alabama Male College located in the town of Auburn, in Lee County. Not only did the buildings and land of the struggling college became part of the new A&M school, its faculty and students remained as well. For the school's first thirty years, it struggled to attract and retain students. At the turn of the century, with no annual state funding, the institution seemed perpetually on verge of collapse. Then in 1907, with news of the boll weevil's devastation in Texas, Louisiana, Arkansas, and Mississippi, the state legislature finally agreed to begin supporting the school and its farm research on a permanent basis. After 1910, as the boll weevil began devouring a sizable chunk of the state's annual cotton crop, legislators became even more convinced of the need to support the college's research and teaching resources.[12]

Twenty miles southwest of Auburn, in the Macon County town of Tuskegee, another A&M school struggled for funding. The Alabama legislature had founded the Tuskegee Normal and Industrial Institute in 1881 as the state's first institution of higher education for African Americans. Like Alabama Polytechnic, the school could not rely on the state for annual funding and depended instead on private donations. In 1895, the fate of the school changed abruptly with founder and principal Booker T. Washington's "Atlanta Compromise" speech, delivered to the Atlanta Cotton States Exposition. The address argued that African Americans should not again press for the "folly" of political gains realized briefly during Reconstruction, but should instead learn to work and "contribute to the markets of the world." Washington's school became the institutional embodiment of his philosophy, and donations began pouring in from northern philanthropists.[13]

An integral part of Washington's philosophy was the improvement of black farm life. When the educator spoke of "industrial education" he included farming; indeed, the school's early years were spent largely in developing its resources for agricultural research and teaching. In 1896, Washington hired George Washington Carver, the first black graduate of Iowa State University. Washington appointed Carver head of the school's agricultural department and the pair immediately began a course both to educate the institute's students and to reach out to the black farmers in rural southeast Alabama. Washington believed that African Americans who paid attention to their growing methods would improve not only their own lives but the lives of their communities as well, and Carver, at least for the first decade he was at Tuskegee, was a committed proponent of the latest scientific methods for increasing crop yields.[14]

Beginning in 1892, Tuskegee held annual farmers' conferences, a day when area landowners and tenants could visit the campus, listen to lectures about farm advances, and ask questions of the school's faculty. As Lu Ann Jones has written, those "who attended annual farmers' conferences at Tuskegee resolved to remove the yoke of the mortgage system . . . and to dedicate themselves to obtaining better schools, churches, teachers and preachers." Resolving to free oneself from tenancy and debt by visiting a Tuskegee class did not, however, easily translate to actual freedom from those forces.[15]

The constant pressures faced by Alabama farmers, both white and black, were rooted in the environmental and economic realities of the cotton fields. The Wiregrass wasn't suited to cotton like the Black Belt was, but even on the best land the international cotton economy kept farmers on the edge of poverty. The professors and scientists at Tuskegee and Alabama Polytechnic sought ways to solve these problems, and they recognized in the boll weevil an opportunity. Each school looked for ways to use the encroaching pest, and the myths that their constituents had heard for years before the insect's arrival, to not only further the public's knowledge of their educational work, but also to gain access to farmers previously reluctant to work with them.

The boll weevil put the flaws of the cotton-based mono-crop system on the front page of every newspaper in the state and the spread of the stories and songs from Texas, Louisiana, and Mississippi made sure that even the most remote farmers believed that the pest was going to change their farming ways. It was the state research and educational agents' job to make farmers connect the presence of the boll weevil to the promise of long-term economic sustainability through crop diversification.

*

The researchers in Auburn and Tuskegee were joined in their effort to educate farmers and to modernize the rural farm by a host of other reformers. The ideology of diversification—the belief that if farmers grew less cotton it would rescue the South not only from the boll weevil but also from intractable rural poverty and tenancy—had found an audience at the highest levels of government. In fact, by the time the pest showed up in Alabama, many administrators in Washington were claiming that the boll weevil had been slowly moving through the South convincing farmers to cease cotton growing.

On May 31, 1907, President Theodore Roosevelt told a Michigan conference of land grant college presidents, "The farmers in the region affected by the boll weevil, in the course of the efforts to fight it, have succeeded in developing a most scientific husbandry." Despite the fact that the weevil was less than halfway across the Cotton Belt and that few signs of a permanent move from cotton existed, the president declared victory. The ravages of the boll weevil had rescued southern cotton farmers from cotton itself. "Not only did the industry of farming become of very much greater economic value in its direct results," Roosevelt told the educators, "but it became immensely more interesting to thousands of families." To hear the president tell it, the boll weevil had moved through Texas and Louisiana leaving not dead cotton and a forlorn people in its wake, but an energized rural population who raised stock and organized corn clubs. Echoing the line spoken so often by progressive farming voices, Roosevelt announced that "in many places the boll weevil became a blessing in disguise."[16]

The southern A&M presidents undoubtedly recognized the spirit of Roosevelt's comments, but whether they considered them a dream or a lie is not clear. The president's remarks spoke to a vision that farm educators had embraced fifteen years earlier when the boll weevil first crossed the Rio Grande, namely that the insect menace would convince farmers to leave cotton for other ventures. It had become a core component of the boll weevil myth. The pest was so devastating, the story went, that farmers had had little choice but to find alternative crops to grow.

Roosevelt's Secretary of Agriculture James Wilson, writing in 1909, echoed the president's misconceived vision that diversified farming had already taken hold in the South. Wilson recalled the earliest days of Knapp's demonstration service, describing how "when the boll weevil came bankers and business men lost confidence and extensive local panics resulted." This was the boll weevil myth at work, and Wilson only propagated it further. He argued that the beetle had brought farmers to the schoolhouse door, so to speak. The secretary bragged of the increase in the number of southern agents from a single person in 1904 to 450 men and women in 1909. "More than 75,000 farmers are receiving direct instruction on their farms," he added.[17] Despite the impressive numbers of agents and the farmers they supposedly helped, Wilson had to admit that efforts to convince the majority of southern farmers of the folly of the one-crop system had actually failed, but he maintained that this insect enemy held the key to making diversification a reality.

Wilson believed the boll weevil had woken southern farmers to the nightmare scenario that cotton farming would no longer be possible, and he knew this in part because he and his agents had helped to spread that idea. Now that his agency, along with local agriculture colleges and state farm agents, had farmers' attention, his task became engineering a viable diversification strategy. "The problem of meeting the advance of the weevil in the South is a complex one," he admitted, but Wilson and the USDA recommended a two-part course of action "in order that the farmers might be able to raise cotton at a profit and in sufficient quantities to meet the world's demands." First, the "adaptation of modern cultural methods" and second, "the teaching of modern farm methods by which *other* standard crops can be produced for the purpose of furnishing food for the family and feed for the stock." The former suggestion was a barely updated version of the cotton cultivation methods Townsend had recommended twenty years earlier. The latter meant growing something other than cotton. It was a recommendation familiar to any farmer that had met with an agent or read a newspaper during the previous decade. It boiled down to a confusing, if not self-contradictory, mantra: better cotton farming and less cotton farming.[18]

The diversification message of Seaman Knapp, Wilson's top extension agent, was no less clear. He encouraged all southern farmers to "raise the food for the family and for the farm stock" as the first priority. Only after a land-owner had devoted sufficient acreage to sustaining the people and animals of the farm should he turn his attention to cotton, "so that his principal cash crop may be all profit." Knapp was convinced of the possibilities of "other" crops in the South. The demonstration chief wrote one Louisiana congress-man that the South was better suited to produce "Indian corn" than the Mid-west. Forage plants like soybeans, velvet beans, and cowpeas were perfectly suited for the region, he argued. And the real money-making promise, Knapp told lawmakers, lay in stock raising. "The South has not developed" its pas-ture lands, he argued "because the farmers have been so engrossed in other crops that they have paid but little attention."[19]

Knapp knew farming well, but it was an unrealistic expectation from a market standpoint. Greater independence from cotton would have meant a fundamental shift in the business of the rural South. The *New York Times* had been right to call the fleece "gold." Cotton was a currency for southern farm-ers. Neither landowners nor tenants could summon credit from thin air; they had to promise the lender that there would be some kind of crop that would pay back the loan at the end of the season. Diversifying one's farm from cot-ton to food crops, stock, and a cash crop still demanded credit, but farmers found few merchants or banks willing to extend money to grow food to feed their families.[20]

Tenants were in an even less realistic position when it came to diversifi-cation. Most landowners told renters and sharecroppers how much of their allotment was to be devoted to cotton, and few gave tenants permission to grow food for their families and animals as the first priority. Furthermore, tenants could not shop around for credit on the open market; they relied on landowners or on a merchant recommended by the landlord to provide their initial funding. If these credit issuers told the tenant what to grow in order to get financing, then the tenant had no choice but to follow these orders. True crop diversification was a noble plan, but it was hard to imagine a place in the South that could realistically pull it off.

Successful diversification could have a chance only in a place that met an unlikely list of requirements. First, cotton's hold on the local economy and lo-cal culture could not be absolute. Second, the soil, climate, and topography would have to support an alternative crop. Third, the local financial power structure, namely bankers and merchants, would have to be behind the initial move away from cotton, so that farmers would have credit and supplies to grow something else. Fourth, farmers would need to gain practical knowledge about growing a different crop or raising stock. Finally, there would have to be a market for the replacement crop. It would do no one any good if farmers in a certain area moved from cotton into a commodity for which there was no sufficient market.

Despite this list of requirements for diversification to be successful, state extension agents continued to try to persuade bankers, merchants, and farmers to support the policy. While the boll weevil's invasion of Alabama dominated local headlines, agents saw their opportunity to make diversification work. John F. Duggar, president of Alabama Polytechnic Institute, kept a large file of clippings about the pest from newspapers across the South. He knew of experts' dire predictions for Alabama's cotton, and in the years leading up to the pest's expected arrival in the state he slowly built a staff of researchers and agents experienced in the boll weevil fight from states to the west. Entomologist Warren Hinds, for instance, before joining the Alabama Experiment Station in 1907, had worked for the U.S. Bureau of Entomology in Texas and Louisiana. B. L. Moss, the newly appointed state agent of the Bureau of Plant Industry also had experience in the boll weevil territory, formerly serving as an agent in southern Mississippi.[21]

Not only had the Alabama extension service employed workers with direct boll weevil experience, they sought information about living with the pest firsthand. Officials in Auburn sent several researchers to Texas to investigate that state's experience with the insect, and their findings—like those of Stone and Fort in Mississippi—endorsed their own interests. Texans were again able to grow cotton in spite of the boll weevil, which local farmers had apparently attributed to the work of that state's extension agents. The Alabama representatives were happy to hear this. They reported that the weevil had been a "blessing in disguise" in Texas, where "the great prosperity of the country" could be traced "to the boll weevil which forced the adoption of the methods advocated by the Farmers' Cooperative Demonstration Work." Good times reigned in weevil-plagued Texas, the agents claimed, because of the local work of farm agents and educators like themselves. "The people have thoroughly recovered from the ravages of the boll weevil," they optimistically reported.[22]

Alabama officials were hopeful that the arrival of the pest would have a similar effect on the cotton growers of their state; they knew years prior to the weevil's arrival that farmers were already eager for their expertise. "The coming of the boll weevil has awakened an intense interest among the cotton planters of the state," the Auburn experiment station's 1911 report found, "and they are now in a position to accept and adopt recommendations for improved methods, such as would not heretofore have appealed to them." That year, crowds of farmers met Alabama Polytechnic president Duggar when he toured the state's black belt to "answer questions directed to bring out something of the boll weevil," as one paper reported. "He received most rapt attention from the farmers."[23]

Indeed, farmers had been eager to learn more about the boll weevil for the years leading up to its arrival. In 1909 alone, entomologist Hinds received 1,200 letters from nervous farmers. He replied to most enquiries with a slim bulletin titled "Facing the Boll Weevil Problem in Alabama." The following year even more letters came from panicked cotton growers, and Hinds, in turn, published three circulars and four bulletins dealing directly with the weevil. He also began traveling the state giving lectures to farmers' institutes and business groups. He and Moss also privately wrote the state's demonstration agents asking them to keep an eye out for the encroaching insect and to send any samples immediately to the state laboratory in Auburn.[24]

These early efforts to distribute bulletins and to meet with farmers to discuss the weevil were as much an exercise in public relations as they were an effort to actually slow the insect's spread or to advance practical solutions. Hinds and his colleagues wanted Alabama farmers to believe in their department's ability to fight the pest and to help their farming generally, but their intention was not to calm the public. More important to the state's agricultural leaders at this stage was positioning themselves at the center of the solution.

For example, in "Heading Off Boll Weevil Panic," a bulletin sent around the state in 1911, Hinds was measured but negative in his appraisal of the weevil's arrival. He guaranteed that the pest would reduce yields at least 25 percent, and advised that nothing short of "immediate diversification of crops" and other changes to the "agricultural and economic systems" would prevent "large loss from the boll weevil." Growers could avoid widespread cotton destruction only with the execution of Hinds's plan.[25]

Hinds's advice was not directed solely at farmers. His article included a pointed warning to "bankers, cotton factors, merchants and others relative to loans or advances." Economic disaster would surely result, he argued, should these groups recall loans or refuse to issue new ones once the boll weevil appeared. Without credit, tenants would simply "be forced to move . . . again starting 'panic.'" It was these business interests' responsibility as "intelligent whites" to direct and help "the blacks." If financial concerns "stand shoulder to shoulder," Hinds wrote, "victory in the fight against the boll weevil will be certain." He implored bankers and merchants to continue to extend credit in spite of the weevil's potential damage, but to use that credit as a lever to force diversification. The entomologist asked creditors to loan money only to farmers who promised to employ his department's recommendations. In addition, lenders should contractually oblige farmers to reduce their cotton acreage. "This will allow him also to raise more food stuffs," Hinds pointed out, "and to adopt some reasonable and profitable diversification and rotation of crops." Hinds used the anxiety created by the pest's appearance as a means to control how farmers farmed; in so doing, he exerted pressure on the state's financial community to play its own part.[26]

By the end of 1911, the boll weevil had invaded close to a third of Alabama, and state researchers' efforts to use the pest to convince the public of the importance of their work was succeeding. Though actual cotton loss because of the insect had been moderate thus far, the state legislature passed two pieces of emergency legislation to fund boll weevil research and education. One funded work at the state experiment station in Auburn. This bill was the first direct government funding of the research farm since 1883. The second act allocated $27,000 to demonstration agents already at work in the state. Though these bills would have a great effect on the research in Auburn and the state agent's ability to travel and meet with cotton farmers, it did nothing to slow the pest or decrease its damage.[27]

*

While officials in Auburn were buoyed by the state's new boll weevil appropriations, in Tuskegee, funding continued to be an obstacle to an effective rural education program. Whereas many of the problems faced by the white extension system were ideological—some farmers resisted the advice of the extension service because they either didn't believe in them or resented government interference—Tuskegee's crisis was more practical. In fact, nearly every aspect of campus life was steeped in ideology. Booker T. Washington had provided the school, its students, and faculty, with an intellectual vision, but it was funding for the practical implementation of that vision that remained the greatest challenge. These everyday concerns over funding meant that Tuskegee's research personnel were, at first, less concerned with the coming boll weevil, and less inclined to try to use its myth as a way in to farmers' lives and decision making, than were their white counterparts.[28]

Whereas white extension agents were able to use the boll weevil to try to convince farmers to grow alternatives to cotton, Carver and Washington had trouble just reaching farmers with their diversification and modernization ideas. The school had been giving out advice to the region's black farmers for years, but these warnings had not affected farmers' methods. As early as 1899, Carver had complained in a Tuskegee bulletin that, despite heavy fertilizer use in the Black Belt and Wiregrass, soils were wearing out an alarming rate. He sent out additional warnings about soil degradation in 1903 and 1905, but few farmers diversified or rotated crops in order to replenish nutrients.[29]

Carver realized that the periodic distribution of a farmer's bulletin simply was not an effective means of reaching growers. Tuskegee's teachers traveled to farms and visited outlying churches in Macon County in an effort to reach farmers with their message. But unlike the state-sponsored extension service in Auburn, Tuskegee had no outside support for this field education, and there was a limited number of outings that its representatives could make. Tuskegee needed an inexpensive way to talk to farmers face-to-face, and they needed outside support to accomplish it.[30]

In 1905, with hopes of finding a more efficient and wide-ranging method of rural education, Washington asked Carver about the feasibility of a mobile farm school—a horse-drawn cart loaded with machines, plants, fertilizers, and literature, which a Tuskegee professor could take out into the countryside and use to teach farmers on their land. Carver liked the idea, drew a sketch of what the proposed wagon might look like, and gave it to Washington. The principal took the drawing north on a fundraising trip, with hopes of finding someone willing to pay for the purchase and rigging of the wagon. In New York, Morris K. Jesup, a well-known banker and philanthropist, promised Washington $500 for the project. On May 24, 1906, the Jesup Movable School made its maiden voyage, with Tuskegee professor George Bridgeforth at the reigns—literally.[31]

By the end of summer, Tuskegee officials claimed the Jesup Wagon had reached more than two thousand people per month. Perhaps more importantly, the mobile school attracted the attention of the South's most important agricultural educator, Seaman Knapp.[32] The following fall, Knapp visited Tuskegee's campus and was so impressed by Tuskegee's work that he offered to make a formal connection between the school and the federal demonstration system. Knapp was barred from using federal funds for black extension, but promised that John D. Rockefeller's philanthropic General Education Board would provide money to hire a full-time black extension agent. Washington could draw money from a separate charity, the Slater Fund, to pay the agent's expenses.[33]

This was a landmark agreement, but its results were anything but clear-cut. Washington and Knapp agreed to hire Thomas Monroe Campbell as the South's first black farm agent. Campbell had grown up in Georgia, worked

FIGURE 11. Thomas M. Campbell speaking to African American extension workers, Dadeville, Alabama. ACES Records, Photographs Box 25, "Negro Extension" subject folder, Auburn University Special Collections, Auburn, Alabama

his way through Tuskegee, and graduated the previous spring. In early 1907, he officially became the school's one-man extension service. The USDA covered $500 of Campbell's salary with General Education Board money and Tuskegee paid him the rest, $340, with Slater funds. The two organizations shared the agent's expenses.[34]

Campbell was charged with taking Tuskegee's message of modernization and diversification to the average small farmer in the counties surrounding the institute. Campbell knocked on farmhouse doors, talked to groups of growers in schoolyards, and held meetings after Sunday church services. He spoke to crowds of white and black rural Alabamans (see Figure 11). mkcheck Often he made presentations in tandem with white extension agents or prominent local white landowners. In 1915, for example, he spoke to a crowd at an African Methodist Episcopal church in Inverness and was joined by the local newspaper editor, a prominent cotton buyer and banker, and a county commissioner.[35]

Despite the crowds and company, Campbell found the work unsparingly depressing. "The average Negro rural home continued to be a dilapidated shack," he wrote later, "in which the living conditions were unspeakable." The conditions of the fields were worse. Campbell found no use for most of the machinery that the Jesup Wagon carried. "I was very seldom able to use the cream separator, the churn or the Babcock testing outfit," he recalled, "because so very few people had cows." There was no point in teaching farmers how to use modern implements that they had no access to. As a result, Campbell adapted his message, speaking on a broad range of farm topics with diversification and self-sufficiency at the heart of the lesson.[36]

Tuskegee's efforts to educate black farmers received a boost in 1914 from the landmark Smith-Lever Act, which appropriated federal funds to state agricultural colleges to fund extension education.[37] The act guaranteed some support of each state's 1890 land grant colleges, which were the segregated schools for African Americans like Tuskegee. Despite its broad vision, the passage of the 1914 bill, like the rural development legislation that had preceded it, made no immediate impact on rural life in Alabama. An army of well-qualified men and women did not immediately appear in rural county seats ready to help farmers. (One Alabama agent recalled that after passage of the Smith-Lever Act he was appointed to serve as Randolph County's extension agent, but that he received no substantial training. "You can understand that I was not a very good county agent," Richmond Bailey admitted, and "nobody else was.") Before the New Deal, the successes of local extension efforts were tied not to public policy, but rather to the work of individual agents.[38]

In addition to Carver and Campbell, a growing number of black agents began working rural southeast Alabama with varying levels of success. As the boll weevil got closer, however, many area agents, both black and white, began to see an uptick in farmers' interest in their work. Unlike Bailey, who was unprepared and had little initial success, M. B. Ivy, a black agent in the Black Belt county of Bullock, reported immediate inroads to his work with the area's black farmers. He wrote a small column for a local newspaper in which he advised farmers to "swat the boll weevil" by deserting cotton immediately and forever. "Destroy all cotton stalks at once," Ivy wrote, and replace the plant with wheat and corn and livestock:

> When every farmer in Bullock county shall eat bread from his own fields and meat from his own pastures, and [be] disturbed by no creditors and enslaved by no debt, shall stand amid his teeming gardens and orchards and vineyards and barnyards, pitching his crops in his own wisdom — then and until then will the farmers of old Bullock be standing on the threshold of progress.

When he visited black growers in Bullock County, Ivy found that some were prepared for the weevil's arrival. Those farmers with the means, the agent reported, had begun growing fruits and vegetables and selling them in the county seat. Farmer William Ousley told Ivy, "We don't go to town on Saturday or any other day unless we carry something to sell."[39]

While Ousley was probably one of the county's few black landowners, Ivy's work with tenant farmers was hindered by the role of the landowner. It was rare that landlords allowed tenants to grow something other than cotton, but Ivy tried to convince owners that those sharecroppers who were able to be self-sufficient first were less likely to migrate. He undoubtedly recounted for landlords the story of one sharecropper, who had grown vegetables in addition to cotton, and sold them directly to the market in town. "I've got plenty to eat and wear a little money in my pocket," he told his extension agent, "if I go off in this fix from Bullock[,] Hell ought to be my home." The tenant had no reason to leave in search of a better deal somewhere else if he wasn't wedded only to the success of the cotton crop. Though Ivy seems to have had some success convincing owners to allow some diversification of tenants' crops, there was no significant economic motive for landlords to comply. And it remained to be seen if, when the boll weevil finally arrived in southeast Alabama, all farmers would be as receptive to diversification rhetoric as Ivy's demonstrators.[40]

By the middle of the 1916 season, county agents began reporting heavy weevil infestations, as well as building momentum for diversification. The Barbour County agent reported, "This year we had our first real experience with the boll weevil . . . so our cotton yield is low." The Geneva County agent wrote to his supervisors, "You will notice that our yield in cotton is very low. This is due to boll weevil and the July flood." His demonstrators "were making a good fight on the weevil," he claimed, but late season rains helped to increase the beetle's population. In Russell County, the agent painted a bleak picture of devastation. "Some of the demonstration fields were entirely destroyed," he testified.[41]

As weevil damage in southeast Alabama increased, some agents made their diversification pleas to bankers and merchants, asking them to be judicious with their loans to farmers who would grow only cotton. But most businesses were directly invested in cotton, either as lenders, buyers, processors, or marketers. As a result, some tried to help farmers fight the weevil more directly. In Opp, a town in Covington County, the local bank offered a reward of twenty cents for every hundred dead boll weevils people brought in.[42] Though this tactic would never work as a means to actually control the pest population, it probably worked as a marketing ploy to convince farmers that the bankers were on their side in this fight against nature.

Once the pest appeared in the area, local extension workers reported an increasing number of farmers showing up to lectures, asking for literature, and volunteering to serve as demonstrators. "The old time idea that any man can raise cotton and so is a farmer has gone," the Covington County agent wrote in 1916, "and the farmer looks up to the educated man as an agricultural leader and comes to the book farmer for advice." As the landowner turns from cotton to another crop, one agent claimed, he "becomes a better diversified farmer than he was a cotton grower, and necessarily a better citizen."[43]

Many of these farmers interested in crops other than cotton asked specifically about peanuts. Even in the Black Belt county of Bullock, peanuts seemed to be threatening cotton as the major farm commodity. "Since the appearance of the boll weevil," the Bullock County agent wrote in his 1915 report, "the farmers have become very much interested in the peanut industry and consequently will grow a very large acreage of this crop next year."[44] Though southeastern Alabama's move from cotton to the peanut was influenced by the long-term work of agricultural agents, bankers, merchants, and researchers, there were smaller local events that shifted farmers' and merchants' perspectives about the suitability of the legumes to the region's culture.

Carver had organized his laboratory research around the principle that an effective diversification program would have to offer a profitable replacement for cotton. In a Tuskegee bulletin he posed, and answered, a rhetorical question common to the time period: "Since the coming of the boll weevil, what is the farmer going to do for a money crop?" His answer was diversification. "There are several crops," Carver argued in his reply, "from which the farmer can realize more money than from cotton; viz., corn, velvet beans, peanuts, sweet potatoes and cow peas." He concentrated on sweet potatoes and peanuts as cotton's replacement crop and attempted to convince locals of the crops' promise in an unlikely place, Tuskegee Institute's dining hall.[45]

Carver's work was part scientific and part public relations. He not only had to make sure that there were uses for these alternative crops, he had to convince merchants and industrialists that they could be turned into cash just as easily as cotton. To that end, Carver invited a group of prominent Macon County businessmen to have dinner at the institute. As Carver later recalled the story, he served them a grand meal of roasted chicken, mashed potatoes, and assorted side dishes. Following the meal, the businessmen praised the student cooks in Tuskegee's kitchen who had turned out the extraordinary feast. It was then that Carver announced to the men that the food they had eaten was not actually chicken and potatoes. It was peanuts. Everything on the menu, Carver claimed, had been made from peanuts. The businessmen were shocked and, as Carver tells it, convinced of the promise of peanuts. The story, even if more apocryphal than real, reveals Carver's understanding of the nonenvironmental aspects successful diversification was dependent upon. The potential profitability of peanuts, a crop that could be used as food for humans, animals, and soil, supposedly persuaded these Macon County businessmen that the legumes might be a better investment than cotton.[46]

*

In no county was there more swift a change from cotton to peanuts than in the Wiregrass's Coffee County. The main business of the county had long been farming, but its population remained sparse until the 1890s. Enterprise, the county's commercial capital, had only 250 residents in 1897, but the following year the railroad and telephone arrived, and the population soared. By 1903, Enterprise claimed 2,750 citizens. The rail line brought people and industry. In 1906, the Enterprise Mills and Novelty Works began operation, making doors, blinds, and other products out of the timber that came into town from the surrounding counties. By 1910, Enterprise was one of only a handful of towns in all of southeast Alabama that the U.S. Census considered urban.[47]

Coffee County's growth corresponded with the arrival of industry, but agriculture was at the heart of its economy. Farmers had dumped fertilizer on their sandy soil for decades, to make it produce cotton, though Coffee County was never as cotton-centric as nearby Black Belt counties. In fact, for a generation before the boll weevil arrived, Coffee County farmers had been experimenting with other crops, including peanuts. The Wiregrass counties of southeast Alabama had led the state's peanut production for years, although it remained a relatively small part of their overall farm production. This changed quickly when the weevil arrived.

On the eve of the beetle's entry into Coffee County, local business leaders and farmers strove to protect the cotton that was central to their economic health. Area leaders called meetings to discuss the boll weevil and invited experts from Auburn to talk about diversification possibilities. Alabama Polytechnic sent their own version of the "movable school" to Enterprise, and farmers gathered around it to examine dead boll weevils and to hear from the experts about early planting, quick-growing seed, and pesticides. The townspeople had attempted to prepare for the pest as most southern counties had, with lots of talk about diversification but little actual movement away from cotton. The boll weevil caused severe damage in 1915 and 1916; local reports estimated crop losses at near 60 percent.[48]

The weevil's destruction didn't alert Coffee County farmers to peanuts' possibilities—they were already aware that the nuts were an alternative to cotton. George Washington Carver and state agents had agitated for the legumes for years, and most Wiregrass farmers were familiar with how to grow them and where to sell them. The boll weevil had added additional cost to cotton farming in a land where it was already expensive to produce the staple due to fertilizer costs, however, and Wiregrass farmers knew that growing cotton in the sandy soil was a precarious endeavor and that peanuts were more suitable to local conditions. In addition to the pull of peanuts, factors in the cotton market pushed Wiregrass farmers away from cotton dependence. Just as the weevils were arriving in the region, peanuts actually became a more secure investment than cotton. In the summer and fall of 1914, war in Europe disrupted both the demand for cotton and safe shipping routes across the Atlantic Ocean. Locally, farmers had received twelve cents per pound for the fiber in June 1914, but only six cents by the end of the year. Farmers suddenly had more than a few reasons to plant peanuts instead of cotton.[49]

Like Carver's peanut dinner, another highly publicized local event in the spring of 1916 gave the final push that many farmers needed to move to peanuts. That year, H. M. Sessions, an Enterprise banker, decided to use his leverage over one local cotton farmer to conduct an experiment of sorts. C. W. Baston reportedly owed Sessions money, though it is unclear why or how much. Using this debt as leverage, the banker instructed Baston to devote 125 acres of farmland to peanuts. Sessions promised to pay Baston as a peanut demonstrator, assuring a price of one dollar per bushel at the end of the season. Baston acquiesced to the demand and grew the legumes. In late October, he picked a huge peanut crop—eight thousand bushels—and transported it into town. Word quickly spread that Baston had delivered the record crop, and keeping to his word, Sessions paid Baston $8,000 for the nuts. That same year cotton farmers in Coffee County harvested a tiny, weevil-ravaged crop.[50]

Baston's peanuts became the stuff of local legend. Area cotton growers immediately turned to Sessions for peanut seed. The banker, in order to meet area farmers' demand for the goobers, resold Baston's entire crop as seed. Though the farmer's success story became instantly famous throughout southeast Alabama, it was not actually too extraordinary. John E. Pittman, Coffee County's extension agent, claimed that dozens of farmers in the county had moved into peanuts from cotton the same year simply because Pittman had suggested it. "The Boll weevil being with us," he wrote at the end of 1916, "caused us to cast about for a crop to partly take the place of cotton so I recommended peanuts." In fact, the same season Baston made his bumper crop, Pittman worked with twenty-five peanut demonstrators who devoted an estimated two-thousand acres to the nuts. But the individual success of these twenty-five peanut growers could never have the effect of Baston's high-profile harvest. The following year, Coffee County farmers turned thousands of acres of land from cotton to peanuts. At least in a limited way, diversification was finally getting its chance in southeastern Alabama.[51]

For all of the years that a chorus of progressive voices, including extension agents, scientists, businesspeople, newspaper editors, and politicians, had called for immediate diversification of Alabama's cotton fields, Coffee County's move to peanuts happened almost overnight. In 1916, Pittman had suggested peanuts to cotton growers and Baston made his bumper crop; by the end of the year, the local agent could report that Coffee County had "become famous as the Hog & peanut County." In 1917, the county produced one million bushels of peanuts, making it the most productive county in the entire nation. As Table 2 shows, over the course of the decennial agricultural censuses, the move from cotton to peanuts was not minor. From 1909 to 1919, peanut acreage increased more than 500 percent, while cotton growers cut their acreage nearly in half.[52]

Not surprisingly, this quick and enormous increase in peanut production glutted the local market. Every farmer who diversified into peanuts was not greeted in the fall with a guaranteed price, as Baston had been the prior year. Local businesspeople, farmers, and extension agents realized that for all the talk about the profitability of diversification, it meant nothing without a market for the crop. It was clear that the farmers of southeast Alabama could make peanuts, but this was only half of the diversification equation. It was not clear what could be done with all of these peanuts.

George Washington Carver had been at work for almost two decades on ways that the legumes could be made into a variety of marketable products. With the advent of peanut production in the southeastern part of the state, Carver again put on a publicity blitz to convince buyers of the nuts' uses. In a set of farmers' bulletins, Carver noted the legumes' high fat and calorie content and devoted entire bulletins to peanut recipes. In one 1915 publication, for instance, Carver listed no fewer than 105 "Ways of Preparing it for Human Consumption," including five for soups alone (creatively named "peanut soup no. 1," "peanut soup no. 2," "Peanut Bisque," "peanut soup no. 4," and "consommé of peanut"). USDA researchers also got in on the

TABLE 2. Agricultural diversification, Coffee County, Alabama, 1909 and 1919.

	1909	1919
cotton acreage	72,535	41,284
cotton bales	25,207	10,729
peanut acreage	8,559	49,393
peanut bushels	173,012	1,204,958
fertilizer cost	$284,471	$488,385

Sources: Thirteenth Census of the United States Taken in the Year 1910, vol. 6: Agriculture, 1909 and 1910 (Washington, D.C.: Government Printing Office, 1913); *Fourteenth Census of the United States Taken in the Year 1920*, vol. 6, part 2: Agriculture (Washington, D.C.: Government Printing Office, 1922).

peanut boosterism. A 1917 *Yearbook* article advertised tasty recipes for "creamed peanuts and rice" and "peanut fondu" [*sic*]. Despite these "appetizing" pleas for farmers to eat more of the nuts, human consumption did not even come close to creating a sufficient demand for the commodity. Peanut growers needed industrial outlets for their crop if the legume was truly going to be a profitable replacement for cotton.[53]

Peanuts had historically been a secondary crop in the South, but as southeastern Alabama began producing more and more of the nuts, farmers found ways to make use the crop central to a profitable operation. Farmers had grown the "goobers" mainly to replenish the soil and feed livestock, particularly hogs, which were fattened for the market by being allowed to graze peanut fields in late fall. With the advent of a serum to control hog cholera in 1906, and the state's funding of a veterinary program to produce the medicine, hog raising became an increasingly safe investment and one that could grow alongside peanut production. The Swift Company recognized this rise in peanuts and hogs in southeast Alabama and built two large meat processing plants in the region. R. C. Conner, owner of a cottonseed oil mill in Enterprise, took advantage of the new local peanut supply and adapted his plant's compresses to process them. He bought the legumes by the truckload, becoming the first peanut oil producer in the county. Sessions himself built a factory to make peanut butter.[54]

With the advent of peanut and pork buyers, the field crop boomed in southeast Alabama. The African American agent M. B. Ivy declared in his 1916 report that "the hog industry is becoming the leading industry of the county." Black and white farmers alike were making "special preparations for the growing of more and better hogs," he wrote. In 1915, no railcars of hogs had been shipped out of Bullock County, but by the following year sixteen cars of hogs were sold to market. The Covington County agent admitted that it was the local demand for hogs created by the arrival of a processing plant to the county, as much as the boll weevil, that drove local farmers to raise pigs. "The packing plant in Andalusia [the Coffee County seat] has done more to get the farmers interested in hog raising than all things combined," the agent reported in 1915, "and within another year or two this county will be full of hogs."[55]

To any observer, the promise of diversification in southeast Alabama seemed to have been realized quickly and dramatically. Extension agents, business owners, and farmers alike all claimed that abandoning cotton had made everyone rich. Peanut growing success stories filled the agents' end-of-year reports. "Since the appearance of the Boll Weevil this crop has done more and made more money for Dale County Farmers than cotton did or any crop ever has," declared an agent in 1919. Cotton was no longer the center of Wiregrass farm production, the agent argued. "Hogs and Peanuts is rightly called the Back Bone of this section."[56]

Agents and newspapers alike painted the material results of the pig and peanut fever as a gold rush. Farmers sold their peanuts and walked into stores to pay their outstanding debts with cash. Peanut-rich landowners made improvements to their farms and homes. Agent R. L. King cited diversification's results in Geneva County: "More and better church and school houses. Rural telephones. Hundreds of Automobiles. Light and Water systems, etc." The black agent for Barbour County claimed that among the rural African Americans he worked with "automobiles are replacing buggies . . . . More farmers paying for and reading newspapers than ever." They were using their peanut money to hire better teachers "who are industrially trained and so train their children" in rural schools. W. M. Welch, the African American agent for Macon County, reported black peanut farmers "buying furniture, rubber tire buggies, automobiles, fine buggy horses." The Henry County agent's end-of-season report recorded success in poetic juxtaposed pairs: "Peanuts & hogs. Bank accounts and Automobiles."[57]

The evidence of southeast Alabama's prosperity on the heels of the boll weevil's arrival and subsequent crop diversification is not limited to the anecdotal stories of state agents (who, it should be noted, saw diversification as a personal and institutional victory). The demographic data bears out drastic changes as well. As Table 3 demonstrates, from 1900 to 1919 farm ownership

TABLE 3. Demographics of Coffee County, Alabama 1900–1919.

	1900	1909	1919
Total farms	2,849	3,925	4,789
Farms run by white farmers	2,464	3,259	3,902
Farms run by "Negro and other nonwhite" farmers	385	665	886
Percentage of land area in farms	76.3	82.5	85.3
Percentage of farm land improved	39.5	51.8	58.7
Average acreage per farm	116.5	91.2	77.3
Value of all farm property	$2,187,785	$6,214,850	$14,770,613
Average value of farm property	$768	$1,583	$3,084
Average value of land per acre	$3.67	$10.62	$24.01

Sources: *Twelfth Census of the United States: 1900* (Washington, D.C.: Government Printing Office, 1902); *Thirteenth Census of the United States Taken in the Year 1910*, vol. 6: Agriculture, 1909 and 1910 (Washington, D.C.: Government Printing Office, 1913); *Fourteenth Census of the United States Taken in the Year 1920*, vol. 6, part 2: Agriculture (Washington, D.C.: Government Printing Office, 1922).

among white and black Alabamans increased, the average value of both farm property and farm land skyrocketed, and the percentage of improved acreage grew.

Wiregrass boosters took advantage of these figures. Geneva County advertised the success of its diversification in the Montgomery *Advertiser*, hoping to court land-hungry farmers. The county invited immigrants, offering "to Share With Them Plenty and Prosperity." The text of the full-page advertisement recalled how many pessimists had predicted "the boll weevil would cause stagnation in agriculture in this section as it had done in the states to the west of Alabama." But, the boosters claimed, "the agricultural wealth of this section has continued uninterruptedly." The prosperity of hogs and nuts had replaced the cyclical poverty of cotton: "Where once broad acres of cotton bloomed, there is now a green carpet of peanuts, dotted with hogs of improved type, gathering their own feed, that will bring the top prices in the markets."[58]

Geneva County's print advertising campaign soon looked diminutive when compared to the boosterism efforts of neighboring Coffee County. Business leaders there were not going to let the quick ascent of peanuts to the fore of their economic life go unnoticed. In 1918, Roscoe Fleming, a merchant, farmer, and member of the Enterprise city council proposed a grand monument to the county's diversification: a statue to the boll weevil. Like the

FIGURE 12. Photograph, "Boll weevil monument, Enterprise, Alabama" (date unknown). The basin at the top of the statue was eventually replaced with a large boll weevil. ACES Records, photographs box 16, "cotton" subject folder, Auburn University, Auburn, Alabama.

boosters in Geneva County, Fleming hoped that erecting a statue to the boll weevil would garner the county favorable publicity, and attract the attention of industry and settlers. He began soliciting contributions from area farmers and businesses and ordered from Italy a stock metal statue in a classical design. When it arrived several months later, however, he had not raised the necessary funds to pay for it and its sixty pieces sat in the Enterprise train station for several weeks until Fleming finally agreed to pay for half of the statue's $1,795 cost himself.[59]

On December 11, 1919, an estimated five thousand people from around the South gathered in a driving rain at Enterprise's central downtown intersection of Main and College Streets, not coincidentally across from Fleming's own store, to witness the unveiling of this monument to the boll weevil. The statue stood in a circular fountain in the dead center of the intersection (Figure 12). Towering thirteen and a half feet tall, it depicted an alabaster-white woman in a wavy dress, with her arms stretched above her head. She held a bronze basin, which spouted water high into the air. A marker unveiled on a nearby corner proclaimed:

> In profound appreciation
> of the Boll Weevil
> and what it has done
> as the Herald of Prosperity
> this monument was erected
> by the Citizens of
> Enterprise, Coffee County,
> Alabama

Slated to speak at the unveiling was George Washington Carver himself. Walter M. Grubbs, president of the Peanut Product Corporation of Birmingham, a company that had bought tons of peanuts from Coffee County farmers, had written to Carver asking him to attend the monument's dedication. "I want to present you, in person, to that town, as I have told them about you in my feeble way," he told Carver. "They want to see you and know you. Will you go?" It was a telling offer. Carver was being asked to share the public stage with white county leaders near the spot where only eighteen years earlier, a white mob had gruesomely and very publicly lynched John Pennington, an African American accused of rape. In spite of this recent history, Grubbs repeated his invitation to Carver, assuring the professor, "We will be the honorees by reason of your presence on the trip. I am looking after your comfort and conveniences at Enterprise. I want you on the stand or platform, with me, or us. I am reserving for myself the pleasure of introducing you." Carver agreed to appear at the dedication—it was certainly not a new experience for the Tuskegee scientist to appear in front of white audiences in sites of previous racial violence—but several days of rains flooded the rail lines from Tuskegee to Enterprise and he was unable to make the trip. In his stead, an agricultural agent of the Southern Railroad, along with several local politicians, addressed the crowd. The statue was, in the words of one observer, "a beacon pointing ever toward the saneness of diversified farming."[60]

The dedication of the boll weevil statue in 1919 marks a remarkable time in the story of southern farming and the boll weevil's profound threat to the social and economic system of the South. The monument declared that the promise of diversification had been fulfilled. Not only did the statue mark the end of King Cotton's grip on southeast Alabama, but it seems that, with the organizers' inviting Carver to make the dedication, a new day of racial harmony had arrived in the Wiregrass. A Montgomery newspaper even went so far as to claim international implications for the Enterprise statue. "Maybe bolshevism is derived from the word bollweevil [*sic*]," the paper ventured. "But we know of nobody who contemplates erecting a monument to bolshevism in Russia," the *Advertiser* argued. "Bolshevism can't bring the practical blessings to Russia that the bollweevil [*sic*] has brought to Coffee County, Alabama."[61]

Perhaps the unveiling of the statue in Enterprise did mark the realization of the diversification dream that so many Americans had had for so many years. And perhaps the lessons of southeast Alabama would spread throughout the South, and others in the boll weevil's path would also turn away from cotton. Industrialists would see the rural South's move away from the mono-crop system and build factories and industries there. For the region's boosters, the unveiling of the boll weevil monument was the beginning of this dream. But if this was to be the story that the boll weevil statue seemed to promise, diversification would have to be more than a fad. It had been a very short time between 1916, when Baston delivered his famous bumper peanut crop to Enterprise, and the unveiling of the statue in 1919. The real question was not how quickly diversification would spread to the rest of the South, but whether it would last even in Coffee County.

"You Will Be Poor and Ignorant and Your Children Will Be the Same": The Boll Weevil Myth Transformed

The dedication of Enterprise's monument to the boll weevil on December 11, 1919, marked the most public celebration of diversification's success in the South. The county's peanut and cotton figures for that year are impressive in their stark contrast. Cotton acreage was down 43 percent since 1909, but over the same period peanut acreage had increased 577 percent. The year Enterprise declared the weevil the "herald of prosperity," peanut acreage actually surpassed cotton acreage in Coffee County. The merchants and peanut brokers who funded the statue's construction believed the monument would serve as a permanent reminder of the folly of cotton and the one-crop system, but in truth, most Coffee County farmers paid no attention to the plaster effigy as they considered what crops to plant the following year.[1]

Despite the boll weevil monument and promising crop figures, when the crowd gathered that rainy December day in Enterprise to see the statue unveiled, it witnessed not the beginning of a broad crop diversification movement, but its brief apex. For a combination of economic and cultural reasons, in the years to come Coffee County farmers disregarded the statue's reminder to diversify. Despite the development of peanut and hog processors, which were supposed to have guaranteed a local market for the legumes, cotton quickly returned as the major economic force in Coffee County and the rest southeast Alabama.

This failure of diversification, and indeed the failure of the boll weevil statue's message itself, offers an opportunity to pull back our interpretive lens, to understand why cotton monoculture was so entrenched in the South and why the USDA and state extension services were so consistently unsuccessful in their efforts to undermine cotton's grasp. The moment of diversification's failure in Enterprise also suggests a shift in the boll weevil myth. When farmers in Alabama continued to grow cotton despite the opportunities offered by peanuts, they demonstrated that they were no longer fearful of the boll weevil. The myth of the pest's all-out destructiveness was no longer convincing. As the boll weevil monument continued to keep watch over downtown Enterprise, the meaning of the pest's myth became less about fear of its wrath and more about the excuse that it offered. Farmers could use the insect as a means to explain their own poverty and slim agricultural options. The boll weevil myth, as a result, began to change from a promise that the weevil would end cotton culture to an explanation for the continuing demise of southern rural life.

*

Before discussing why farmers turned so quickly back to cotton, it is important to look at the specifics of the crop's swift rebound. How, where, when, and to what extent the fiber returned as the region's principal crop reveals a great deal about the limits of the diversification plan and of the boll weevil monument's claim. As Figure 13 starkly portrays, for the first five years after the statue's dedication, peanut acreage in Coffee County climbed only marginally, but cotton's return was dramatic. Only a couple of years after they had announced their allegiance to the peanut, farmers around Enterprise again planted more land in cotton than in the legumes. Ten years after the statue was erected, cotton acreage had risen to pre–boll weevil (and pre-statue) levels. Cotton returned to Coffee County almost as quickly as it had gone.[2]

Not only in the statue's home county but also across southeast Alabama, farmers had given diversification a try, and almost without exception they returned to cotton in the years that followed. All counties in southeastern Alabama were not equal in their commitment to cotton, however. In 1929, farmers statewide devoted 43 percent of their cropland to cotton, but among southeastern counties, those lying predominantly within the Wiregrass planted less cotton than the state average. By the end of the 1920s, farmers in Black Belt counties like Macon relied on the white staple to a greater degree than the state average, but in the Wiregrass, where cotton was more expensive to produce, only a little evidence of the peanut movement remained.[3]

More important than acreage, however, was income. According to one study, Alabama farmers derived 72 percent of their income from cotton. However, as Table 4 demonstrates, in Black Belt Counties this rate was

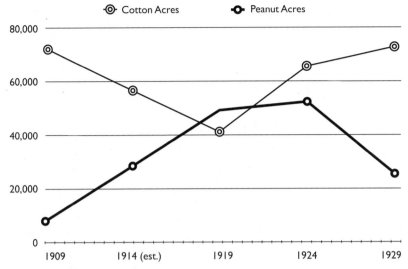

FIGURE 13. Coffee County, Alabama, cotton and peanut acreage, 1909–1929. Data from *Thirteenth Census of the United States Taken in the Year 1910*, vol. 6: Agriculture, 1909 and 1910 (Washington, D.C.: Government Printing Office, 1913); *Fourteenth Census of the United States Taken in the Year 1920*, vol. 6, part 2: Agriculture (Washington, D.C.: Government Printing Office, 1922); Agricultural Census, 1925 (Washington D.C., Government Printing Office, 1927); *Fifteenth Census of the United States Taken in the Year 1930*, agriculture vol. (Washington, D.C.: Government Printing Office, 1932).

significantly higher. In the Wiregrass counties of Pike, Dale, Henry, and Coffee, farmers earned more than 16 percent of their crop income from something other than cotton. But even Coffee County's seemingly impressive figure of 21 percent noncotton crop income paled in comparison to the numbers a decade earlier, when diversification had been given its chance. By 1930, cotton still accounted for more than 66 percent of the county's farm income. In short, despite diversification schemes, Black Belt farmers turned to their rich soil for cotton, aided by an increasing amount of fertilizer. In the Wiregrass, farmers were only slightly more willing to plant peanuts and sweet potatoes, and to raise livestock, and these alternatives never amounted to much of a threat to cotton.

Though these figures reveal that the statue's claim of successful diversification was at best short-lived, what of the monument's assertion that the boll weevil had been the "herald of prosperity"? Simply looking at short- or long-term diversification figures cannot test economic success. Perhaps the return to cotton was accompanied by a decrease in tenancy or poverty.

TABLE 4. Farming sectors as percentage of the value of all farm production, southeast Alabama, 1929.

	All field crops	Cotton (including seed)	Crops other than cotton	Livestock	Livestock products	Forest products
Statewide county	81.98	72.54	9.44	6.31	9.72	1.99
Macon	84.45	78.80	5.65	5.59	8.12	1.84
Russell	79.90	72.60	7.30	7.82	7.76	4.52
Bullock	71.99	64.35	7.64	16.85	8.36	2.80
Lee	78.54	70.64	7.90	5.71	12.51	3.24
Crenshaw	85.92	75.96	9.96	6.67	4.85	2.56
Houston	88.44	78.15	10.29	5.31	5.57	0.68
Covington	82.49	71.31	11.18	7.38	6.99	3.14
Barbour	88.46	76.55	11.91	4.60	5.23	1.71
Geneva	82.26	69.07	13.19	10.82	4.87	2.05
Pike	82.82	66.41	16.41	9.14	5.02	3.02
Dale	81.52	62.16	19.36	9.58	4.75	4.15
Henry	89.88	69.99	19.89	5.18	3.43	1.51
Coffee	87.69	66.56	21.13	6.77	4.15	1.39

Source: W. M. Adamson, *Income in Counties of Alabama, 1929 and 1935* (University, Alabama: Bureau of Business Research, University of Alabama, 1939)

Here again the statue's message rings hollow. Over the entire period from 1900 to 1930, which begins prior to the boll weevil's entry in the county, including its initial appearance and the subsequent diversification experiment, and ends long after most farmers had returned their fields to cotton, it is clear that the weevil did not significantly reduce poverty in Coffee County. As Table 5 demonstrates, the actual picture of "prosperity" claimed by the monument was muddy for both landowners and tenants.

For landowners, the value of farmland per acre rose until 1920, then evened off and dropped slightly before 1930. The value of their farm property similarly peaked in 1920, dropping nearly 30 percent by the end of the decade.[4] This drop in value was clearly bad for landowners, but it could be suggested that lower land prices opened up ownership opportunities to tenants. (Perhaps the prosperity to which the statue's marker referred was tenants becoming landowners.) The figures do not support this notion either.

Tenancy rates soared over the period. In 1930, tenants in Coffee County operated two and a third times as many farms as they had in 1900. Rather than making land affordable and thereby enabling tenants to buy land, the drop in

TABLE 5. Testing the prosperity thesis, Coffee County, Alabama, 1900–1930.

	1900	1910	1920	1925	1930
White-operated farms	2,464	3,259	3,902	4,134	3,442
Negro- and other nonwhite-operated farms	385	665	886	866	732
Total farms	2,849	3,925	4,789	5,000	4,174
Percentage of land area in farms	N/A	82.5	85.3	77.7	77.2
Average farm acreage	116.5	91.2	77.3	67.4	80.3
Value of all farm property ($ millions)	2.19	6.21	14.77	12.82	9.58
Average value of farm property ($)	768	1,583	3,084	2,564	2,295
Average value of land per acre ($)	3.67	10.62	24.01	24.47	19.77
Percentage of tenancy	45.2	57	64.6	69.1	71.3
Farms operated by tenants	1,289	2,238	3,095	3,455	2,982
Total fertilizer cost ($)	N/A	284,471	488,385	611,661	N/A

N/A: not available.
Source: *Fifteenth Census of the United States Taken in the Year 1930*, agriculture vol. (Washington, D.C.: Government Printing Office, 1932).

value seems to reflect the fact that fewer people could afford to buy land or to keep it. By 1930, nearly three-quarters of the county's farmers were tenants. As Kathryn Holland Braund argues, tenancy increases may have been attributable to an influx of poor farmers from other places, courted by the now famous wealth of the peanut farmers in the county. Nevertheless, these immigrants still found themselves poor and landless. Only making matters worse, there was a sharp rise in the cost of fertilizer on which Coffee County's growers increasingly relied to make the soil produce cotton. For tenants, it was just one more expense for which they had to borrow money.[5]

The figures also reveal a more basic change in Coffee County landowner-ship. From 1900 to 1930, the total number of farms rose 46 percent, from 2,849 to 4,174, but the average farm size dropped. Additionally, the total amount of land in agricultural production in the county shrank. In other words, more farmers were farming on less land in 1930 than in 1900, and more were doing so as tenants. As banks foreclosed on landowners unable to meet their debts, a surplus of available land drove down prices. Some of the farms were divided into smaller plots while some land was simply abandoned. Some of the acre-age never made it back into agricultural production at all. Despite the claims of Enterprise business leaders that the statue represented the "prosperity" brought to the county by the weevil, few Coffee County farmers could have been labeled prosperous. Whether landowners, renters, or sharecroppers, farmers were spending more money on less land and trying to keep it productive with increasing amounts of fertilizer.[6]

The remarkable statistics in the tables cannot portray the frustration felt by the region's extension agents. Evidence of diversification's fast rise and fall rang loudly throughout the writings of the agents who had put so much hope in communities' moves away from the mono-crop system. Not only had the educators pushed peanuts and hogs, but they had reported to their superiors that cotton was disappearing from their territories. They had also predicted that along with cotton, tenancy and poverty were departing as well. After 1920, however, they began to realize that tenancy rates were actually climbing, as farmers failed to continue diversification plans. For dozens of southeast Alabama agents, it was a crushing blow.

No one communicated this disappointment better than Covington County agent J. P. Wilson. Less than two months after the dedication of the Enterprise statue, Wilson was distraught to find farmers returning the majority of their fields to cotton. They had dutifully replaced the crop with peanuts and hogs for several years and had success, but it seems that the absence of cotton had fooled farmers into thinking that the boll weevil was no longer a problem. "We have had two right good cotton years with very little weevil damage and abnormally high prices," he admitted, but "these facts have almost completely covered up the results of 1915 and 1916 in the farmers [*sic*] eyes and a large acreage in cotton is being planned for this year."[7]

Wilson begged farmers to reconsider their choice and to look at cotton's effect not just on their bank accounts, but on their lives in the long term. "Let me give you something to think over," he wrote in 1919:

> If you plant ten to twelve acres in cotton this year and the weevils come, as they are almost sure to, and you make about one and a half bales per plow, with nothing else to sell, where will you be next fall? Why not use some horse sense and plant several money crops[?] [W]ill the farmers of this County go crazy again and act like a bunch of small school boys?

He named peanuts, oats, and potatoes as alternatives, and encouraged cow and hog raising. "Help the wife market the eggs and chickens and look after the pigs," he implored. It was near the end of his report that Wilson's plea reached a fevered pitch. In an almost hysterical tone, he promised farmers who continued with cotton that they were inviting devastating boll weevil

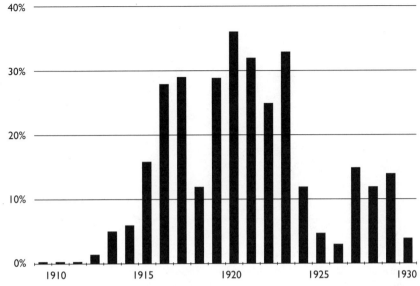

F I G U R E 1 4 . Estimated annual percentage cotton crop loss from boll weevil in Alabama, 1909–1930. Data from Willard A. Dickerson et al., eds., *Boll Weevil Eradication in the United States through 1999*, Cotton Foundation Reference Book Series, no. 6 (Memphis: Cotton Foundation Publisher, 2001), 614–15.

damage, which he promised would return them to intractable poverty and despair. His closing line was directed at these farmers who had returned to cotton. "My friends," he told them, "you will be poor and ignorant and your children will be the same."[8]

Unfortunately for thousands of southeast Alabama farmers, Wilson's prediction that growing cotton meant enduring poverty came true. As growers chose cotton over alternative crops season after season, the boll weevil returned in dreadful numbers. Statewide data show that the insect continued to be a major consumer of cotton in Alabama throughout the 1920s (Figure 14). This return of the boll weevil coincided with the profound failure of diversification in southeast Alabama. What had once been a model for the possibility for diversification created by the cooperation of business leaders, state and federal extension agents, and farmers had become just another cotton-drunk corner of the South.

But why? Considering the power of the boll weevil myth, the reality of its disruption of the cotton economy, and the cycles of debt that cotton carried with it, why did Alabama farmers return to the crop? Put simply, three factors

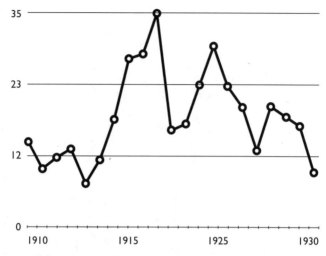

FIGURE 15. Alabama cotton price, cents per pound, 1910–1930. Data from USDA Bureau of Agricultural Economics, "Statistics on Cotton and Related Data," Statistical Bulletin 99 (June 1951), 34.

killed diversification's chances in southeastern Alabama: economics, cotton's cultural hold, and racism. The boll weevil and its changing myth were woven into each of these causes, just as it had been in diversification's rise.

The most important and immediate reason farmers returned to cotton lies in the complicated economy of its production. Demand for the staple had been low on the brink of World War I, and the consequent drop in prices was one reason farmers had turned to peanuts, but as Figure 15 demonstrates, shortages around the world drove prices sky high at the end of the 1910s. In 1914, farmers had been willing to listen to talk of decreasing cotton acreage while prices hovered under ten cents per pound, but in 1919, despite the money made in peanuts, they stood to make a much greater profit by returning to cotton when the price topped thirty-four cents per pound.

Alabama growers embraced the roller coaster ride of prices for a multitude of reasons, but the main factor was that cotton *could* pay off. If a farmer fertilized, planted a fast-growing variety, and fought the boll weevil with poison, he or she could, weather permitting, make a big crop. If farmers then sold their cotton when prices were high, they stood to make a substantial profit. Cotton *could* bring riches to farmers, even considering the boll weevil, if everything fell into place, but it took an amazing combination of factors. Making cotton profitably under weevil conditions meant overcoming a lot of "ifs," but farmers had a hard time ignoring the lure of high prices.[9]

T A B L E 6 . Potential value of cotton versus peanuts, Coffee County, Alabama.

	1909	1919	1924	1929
cotton acreage	72,535	41,284	65,744	72,869
cotton bales	25,207	10,729	22,535	19,198
cotton bales per acre	0.348	0.260	0.343	0.263
cotton price per bale ($)	69.90	174.50	115.40	82.75
dollar value cotton per acre	**24.33**	**45.37**	**39.58**	**21.80**
peanut acreage	8,559	49,393	52,087	25,804
peanut bushels	173,012	1,204,958	1,346,371	966,288
peanut bushels per acre	20.21	24.40	25.85	37.45
Peanut price, cents per pound	3.9	8.0	4.8	3.8
dollar value peanuts per acre	**17.34**	**42.94**	**27.30**	**31.31**

Source: *Thirteenth Census of the United States Taken in the Year 1910*, vol. 6: Agriculture, 1909 and 1910 (Washington, D.C.: Government Printing Office, 1913); *Fourteenth Census of the United States Taken in the Year 1920*, vol 6t 2: Agriculture (Washington, D.C.: Government Printing Office, 1922); Agricultural Census, 1925 (Washington D.C., Government Printing Office, 1927); *Fifteenth Census of the United States Taken in the Year 1930*, agriculture vol. (Washington, D.C.: Government Printing Office, 1932).

In the Wiregrass, where cotton farming was more expensive due to fertilizer costs, the boll weevil had made peanuts a safer economic choice, but rarely a more potentially profitable one. "Even under our conditions with weevils," county agent Witherington wrote in 1919, "figures show some acres yielding a profit of sixty dollars." Though he concluded that "you can't find any crop we can grow that will return more profit than our cotton crop," in truth farmers rarely made sixty dollars per acre on either cotton or peanuts. Taking the census years as snapshots, statistics show that cotton remained a better profit maker than peanuts for Coffee County growers most years, though as farmers gained more experience with the nuts, the gap decreased. As Table 6 attests, the relative prices of peanuts and cotton were the most important factors in determining dollar value per acre of each crop.[10]

The point that Witherington and his colleagues made to farmers over and over was that this per acre value of cotton versus peanuts did not include costs, and that—thanks to the boll weevil—growing cotton was increasingly expensive. Insecticide prices rose as farmers turned back to cotton. "We carried out the calcium arsenate test on one co-operators [*sic*] farm but it did not prove economical this year," Witherington reported at the end of 1919. Farmers had no choice whether or not to apply poisons, however; if they chose not to poison, the weevil would destroy vast quantities of the fiber. If they did apply the pesticide, as Witherington reported, cotton profits did not meet costs.[11]

This kind of hard data, however, rarely swayed the region's small and mid-sized landowners. For cotton farmers across the South, but particularly in places like the Wiregrass, where the soil was less rich than in the Delta and Black Belt, cotton farming was an unpredictable business. Prices fell, weevils appeared in varying degrees, weather helped or hindered the crop — these were all factors that could determine whether making a profit with the white staple was even possible. At the beginning of the season, however, most growers simply eyed the cotton price and decided what crops to plant. Unfortunately, the price at harvest time often had little resemblance to its spring counterpart.

The vagaries of the market and the credit systems were not the only factors in cotton's failure in southeast Alabama. Cotton had a cultural hold on the region that peanuts could never combat. Though the supporters of the boll weevil monument hoped that its presence in downtown Enterprise would entrench peanuts' own social and ideological presence in the region, these men and women failed to realize that cotton had its own monuments. Everywhere one looked in the rural South there were effigies built to the tangible prosperity that cotton could bring. Enormous mansions from the ante- and postbellum periods dotted the southern countryside, serving as their own statues *for* the mono-crop cotton system. The big automobiles that lined downtown streets after a harvest reminded poor farmers of cotton's possibilities. In fact, cotton was such a part of the fabric, the material conditions of rural Alabama, that any sign of wealth was a sign of cotton's possibility. The heyday of peanuts was so short that it never had the chance to carve out its own meaning in the culture to rival that of cotton.

Beyond these economic and cultural factors in diversification's failure, practical systemic problems within the rural education system — most importantly white supremacy — were also key.[12] Racism was, if not a policy of the extension system, an unwritten organizing principle. White supremacy ideologically guided its work, and Jim Crow social restrictions limited the work on a practical level. From the very top of the extension system to the bottom, racism reigned. Seaman Knapp told University of Georgia chancellor David Barrow that despite the boll weevil, the South would prevail as the country's greatest agricultural land not only because of the "germinating power" of its soil, but because its "people are the purest Anglo-Saxon." "To me the Southern people are the purest stock of the greatest race the world has produced," Knapp told Barrow. Beyond the racial assumptions of Knapp's comment, the more telling aspect is that it reveals that he did not consider black southerners a part of the southern people. He surely knew that over one-third of southern farmers were not "Anglo-Saxon," but he did not consider those who were not white or did not own land to be worthy of his tally.[13] At the root of the USDA's and the extension service's rural education philosophy was a contradiction related to racism: most educators believed that tenants and small farmers could not learn and apply the very instruction that agents provided. The majority of local extension agents, and the bureaucrats who directed them at the state and federal levels, believed that the modern farm methods they sanctioned could not be put in place by tenant farmers, white or black, or even by black landowners. From the beginning of the boll weevil's invasion of the South, in fact, experts predicted that the pest would exact the heaviest damage in areas heavily populated by African Americans with high tenancy rates. G. A. Rich, white agent for Bullock County, reported in 1911, prior to the pest's entry to his county, that it "has been apprehended that the boll weevil will do more damage [here] than in the sections where the small white farmer gives his personal supervision to the farm." Alabama Polytechnic's John Duggar warned large white landowners not to leave the boll weevil fight to tenants. "The only feasible thing for large farmers with negro tenants to do," Duggar told a gathering prior to the weevil's arrival, was "to reduce the acreage one-half to the tenant." Limiting the amount of land grown by the "negro tenant" he added, afforded "an opportunity to kill the boll weevils." In Alabama, as had been the case in Texas, Louisiana, and Mississippi, fighting the boll weevil meant dealing first with the nonwhite farm population.

Evidence of this white supremacy runs throughout the records of the extension service. Seaman Knapp, for instance, believed that rural African Americans simply would not understand the diversification message itself. "When I talk to a negro citizen I never talk about the better civilization" he said, "but about a better chicken, a better pig, a white-washed house."[14] Warren Hinds, Alabama's state entomologist, recorded the results of one experiment with a mechanical poison duster, noting that "Machine trouble" was "nothing serious," but "Nigger trouble is worse than engine trouble."[15]

Plucking racist quotes from the record of white southerners in the first-half of the twentieth century is not difficult, although finding white supremacy at every level of the extension service and USDA reveals how racism effected farmers' fight against the boll weevil. If reformers and educators closed certain options to certain people, which they did in the case of black tenants and landowners, it had tangible agricultural and economic results. The health of more than the crop was at stake.

In hidden, sinister ways, white agents' disregard for black farmers caused physical injury and perhaps even death. Auburn's Hinds methodically recorded the results of two experiments with Paris green, an anti-boll weevil arsenical compound applied by laborers to infested cotton. In a multicounty study, Hinds found that the poison was causing poor health in sharecroppers and livestock across the state. His report made little delineation between livestock and black tenants, and made no mention of poisons ever affecting white farmers. In Baldwin County, for instance, a cow had died from Paris green; in Bullock County the poison had caused sores on livestock. But in Dallas, Perry, and Pike Counties, Hinds noted "sores on men and mules," and in Mobile there were "2 negroes sick." "One family [of] negroes" poisoned from the insecticide in Geneva County were listed as "all recovered," but in the other cases Hinds made no notation of the final outcome. The evidence offered by Hinds' chart, found tucked away inside his personal papers—not published in a state bulletin or local newspaper—raises the question of what the effect of the extension-prescribed poison had when scientists were not keeping a tally. At the very least it may explain tenants' own apprehension in working with agents. Why would a farmer literally getting poisoned by extension service advice be willing to work with the agents?[16]

This racism started at the top of the extension chain of command, but its most sinister manifestation was at the county level. Beginning in 1918, extension agents found an additional two-page section, titled "Special Report By White Agents on Work With Negro Farmers," attached to the reports they were required to complete at the end of each season. Despite the new form, which asked agents to specifically record their work with black Alabamans, the overwhelming majority of white agents simply left the pages blank, indicating either that they had no contact with black farmers or that this work was not important enough to report.[17]

Those agents who did fill out the form have left a telling document on the extension service's work with black farmers. Filling out the form for the first time, the agent for Covington County wrote simply, "We have no negrow [*sic*] farming section in the county," explaining that "the Negrows all farm for or with white farmers." According to the census, however, Covington had nearly 8,000 black farmers in 1920, including 165 landowners. S. N. Crosby of Dale County claimed there were "very few Negro farmers in this county" despite the fact that over half its farms were operated by African Americans, some 5,000 men and women. In 1920, Crenshaw County's agent claimed that "there are very few Negro farmers in this county ap[p]rox. 95% of the farmers are white." The census reported less than 65 percent of Crenshaw's residents were white; the figure was probably even lower for the farming population. In Bullock County, where 81 percent of farmers were African American, the white agent noted simply "I have left this part of the work to our negro county agent."[18]

Other white agents were more realistic in their explanations of why their work centered on white landowners. "Among the negro farmers in Geneva county there is scarcely a land owner," agent Rufus King wrote, "consequently the farms operated by the negro farmers are done under the supervision of the white landlord." King worked only with those black tenants whose landlords allowed their contact, and found that "the negro farmers in the county . . . are no less responsive to instruction and encouragement than are white farmers of the less intelligent class." Non-landholding whites, in King's view, were no better or worse equipped to receive his teachings than black tenants. With no sense of irony, King reported the following year that black farmers were contacted only in connection with the efforts to educate white landowners, but that "no discrimination is made of the two wherever service can be rendered." Macon County's white agent claimed to "answer Negro calls as quickly as I do White calls." "To a man they always follow Boll Weevil instructions," he wrote at the close of 1918. "We have proven that cotton can be raised under boll-weevil conditions."[19]

White agents were clearly working within a racist system. The white supremacy of landowners, merchants, bankers, and politicians was no less important than agents' own racism when it came to the failure of extension work. Agents were pressed with making rural Alabama modern and diversified, but their hands were tied when it came to reforming issues most pertinent to rural African Americans, namely access to landownership. Clearly from the above testimonies, however, white agents rarely did anything other than endorse the racism at the heart of the rural economic and social structures. Black agents were rarely more successful at truly reforming African American rural life.

Black extension agents faced an even greater set of problems in executing federal education strategies than their white counterparts. Thomas M. Campbell of Tuskegee worked tirelessly with farmers white and black, but his effort showed few signs of success. For all of the fame surrounding the Jesup Wagon and its successor, the Booker T. Washington Movable School, the mere ability to travel into remote areas and teach farmers did not automatically mean that rural growers modernized their operations and diversified their fields. Even black extension agents like Campbell could not overcome the restrictions of tenancy and racism.

Few black or white agents worked directly with tenants, and as Table 1 shows, the Black Belt counties surrounding Tuskegee had high African American populations, but few black landowners. As a result, Campbell and his colleagues were forced to first approach the most powerful white landowners in order to access tenants. Even then, pleas for diversification directed to tenants proved meaningless; only landowners made the decisions of what, how, and when to plant. As Mark Hersey has shown, over time George Washington Carver developed a system that would help poor farmers, one that took advantage of the local bounty offered in the Black Belt's forests and fields, but Jim Crow stood in the way of any widespread implementation of Carver's plan.[20]

Black extension agents were left doling out advice about lifestyle and consumption to tenants and small landowners. The black agent for Barbour County echoed Booker T. Washington's message in 1918, telling farmers "'Let your bucket down where you are.' Buy your farms and settle down," but the advice carried no concrete economic plan for tenants to make economic uplift a reality. He did offer three pieces of advice for black tenants in search of a better life on the farm. First, "kill out all of the sorry, worthless dogs." Second, "stop free feeding strong, sorry worthless people." And finally, if killing domestic animals and pushing out people who would not work hard did not solve their problems, black Alabamans were advised to "stop buying lots of *cheap* jewelry, organs, mules, machines and other things generally brought by peddlers."[21]

Some scholars have even argued that black tenants and landowners in southeast Alabama actively refused to work with black agents from Tuskegee or the USDA. Karen Ferguson suggests that to many black farmers, black agents represented accommodation, not uplift. The substance of the modernization message, some tenants believed, was one intended to push African Americans toward white control. Black farmers may have recognized that the ideas behind boll weevil control, diversification, and modernization were intended not to guide black farmers to independence, but to bring black farmers into the white-dominated rural economic market. Whether Ferguson's claim is true or not, scholars have agreed that throughout the first decade of the boll weevil's appearance in Alabama, black extension agents' efforts to reform farming rarely worked.[22]

The success or failure of diversification hinged in particular instances on this agent-farmer relationship, but fundamental economic forces almost always worked against diversification. In places where agents worked with small farmers there was some evidence of cotton loosening its grip on rural life, but in most places the breakdown of the agent-farmer relationship meant the final end of the diversification dream as well. This was clearly not only the result of farmers refusing to work with agents, or educators' elitism when dealing with farmers. Within the economic, social, and cultural system of agricultural production in the rural South there was little room for any kind of talk of true reform of systems. Within these constraints, agents did what they could do to help farmers and, in turn landowners and tenants alike tried to retain more cash and personal independence.

Diversification's failure in Alabama—in the very place where the most famous monument to this idea still stands—was swift and profoundly revealing. Though high cotton prices had been an initial impetus for farmers to stop their peanut and hog production, the failure of the relationship between farmers and the extension service, and particularly the role racism played therein, contributed to the long-term death of the diversification dream. As a result, poverty rose dramatically. By 1930, per capita income in these southeastern counties trailed other parts of the state, a state that was itself near the bottom of the national income average.[23]

CHAPTER 7

Cotton's Obituaries:
The Boll Weevil in Georgia

A history of Washington County, Georgia, published in 1989 states that, when the boll weevil appeared there shortly after World War I, "hordes of devastation covered the countryside." By the 1920s, according to the book, the insect pest "was desimating [*sic*] the county's cotton crop." Despite this bleak picture, the author admits that by the end of the decade, "few farms were lost, few businesses failed." In an almost apologetic tone, the writer suggests that Washington County citizens felt few effects from the pest, "possibly because by nature its people had been conditioned to be conservative and parsimonious." The interpretation attempts at once to endorse the myth of the weevil's destructiveness and to argue that the pest actually had only a small local effect. This contradiction speaks to the power assigned the boll weevil at both the beginning and end of the twentieth century. For generations, Georgians have encountered this historical narrative that the boll weevil ended the state's allegiance to King Cotton. It's a story soaked into the fabric of rural Georgia's histories, stories, songs, and most importantly its institutions. A close examination of the record reveals, however, that this explanation itself—the argument that the boll weevil ended farmers' dependence on the crop—was the primary factor in Georgia farmers breaking their allegiance to the fiber.[1]

The image of the boll weevil as the destroyer of the plantation system has proven stronger and more lasting in Georgia than in any other state. It is an impression nurtured prior to the weevil's arrival by the state's own agricultural leaders, both in the public and private sectors, and fostered by a cross-section of Georgia society after the pest actually began destroying the state's cotton. While planters sought a more favorable credit system and greater restraints on labor, and used the weevil to try to convince legislators to make bills favorable to their class, farm educators were the main force behind making the weevil a statewide concern. To present a ground-level view of how the weevil became a scare tactic aimed at the state's mid-sized and small farmers, this chapter focuses specifically on two individuals within the state's extension service bureaucracy. Many landowners and tenants, in return, used the boll weevil as a specific, material reason to quit cotton and to leave rural Georgia altogether. Finally, the most important factor in the enduring myth of the boll weevil's effect on Georgia has been the work of academics who studied the state during a time of a major outmigration by thousands of the state's rural cotton laborers. This chapter concludes by revisiting the role that artists and singers had in codifying the boll weevil myth in songs whose relevance lasted throughout the twentieth century.

*

Despite the claims of Georgians — farmers, politicians, extension agents, writers, musicians, and scholars — the boll weevil's invasion did not mark what sociologist Arthur Raper has called a "preface to peasantry." In fact, the majority of the state's rural residents had been stuck in a condition nearing peasantry for generations prior to the pest's appearance. In the forty years leading up to the pest's 1915 arrival in the state, no visitor to rural Georgia could have detected a marked improvement in the quality of life of most landless farmers. Why then, if the boll weevil didn't itself fundamentally disrupt the rural economy of Georgia, has the destructive image of the pest held for so long?

The answer lies in a series of economic and social changes in rural Georgia that occurred around the time of the weevil's appearance. Indeed, there was real change in the state that corresponded to the boll weevil's destruction of Georgia cotton. The first two decades of the pest's presence in Georgia coincided with a slow breakup of the cotton plantation system in parts of the state and a massive migration of the rural workforce, but the cotton boll weevil was hardly the principal factor in this demise. The destruction of cotton exacted by the insect, though devastating for many individual farmers, was merely the proverbial straw that broke the camel's back, ending cotton production that had already grown too expensive on marginal farmland throughout the state. Rather than a "preface to peasantry," the boll weevil's entry to Georgia marked its postscript.[2]

An observer in the 1910s, however, could easily have been swayed by the

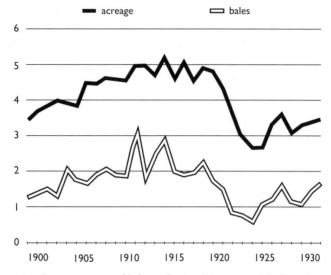

FIGURE 16. Cotton acreage and bale production (× 1,000,000) in Georgia, 1900–1930. Data from Willard A. Dickerson et al., eds., *Boll Weevil Eradication in the United States through 1999*, Cotton Foundation Reference Book Series, no. 6 (Memphis: Cotton Foundation Publisher, 2001), 598, 601, 604.

illusion that Georgia cotton farmers were enjoying a period of great prosperity. As Figure 16 demonstrates, there was a marked rise in both cotton acreage and bales produced in the state from 1900 to 1917. As in southeast Alabama at the same time, farmers in Georgia had responded to rising cotton prices during World War I by planting more of their fields in the white staple. The state's gins and ports filled with record numbers of cotton bales. Newspapers heralded the crop's upsurge as county cotton production records fell year after year. It seemed to most observers that the pessimists who had predicted cotton's demise in the 1890s, promising only that diversification could save farms from bankruptcy, had been wide of the mark.[3]

A close examination of the figures, however, reveals that despite higher prices and production, cotton farmers in this so-called Golden Era were in fact falling deeper and deeper into debt. For the decade prior to the boll weevil's invasion, cotton production rose alongside prices, but farmers' profits did not keep pace. The costs of land, seed, fertilizer, and ginning all increased as well, making it as hard to earn a profit when cotton was selling for thirty cents per pound as when the price was near eight cents.

Two studies by the state's College of Agriculture and the USDA's Bureau of Farm Management in 1913 and 1918 make this point in painful detail. These reports balanced the average cost of production against the price farmers received. In 1913, when cotton sold for 12.23 cents, Georgia farmers needed 16.00 cents per pound to break even. Five years later, cotton prices had climbed to 23.39 cents per pound, but the price growers needed to break even had soared to 32.00 cents per pound. At that price, only 56 percent of farmers managed any profit. Cotton farmers knew what many outside observers did not, that the rise in prices had not meant a rise in profits. It was a complicated argument to make, but luckily for farmers and reformers in Georgia, a ready-made excuse for growing debt and poverty was on its way.[4]

The boll weevil first entered the state in late 1914. By the end of the cotton season the following year, the pest had traveled nearly half the distance across Georgia. In 1916, farmers in every cotton-growing corner of the state could find weevils in their fields, but it did not begin to seriously threaten the state's crop until 1919, just about the time cotton prices in the state plummeted (see Figures 17 and 18).[5] The pest increased farmers' costs substantially as it forced those serious about their efforts to fight the bug to purchase huge amounts of poisons, extra fertilizer, more expensive seed, and other supplies. In other words, farms that were struggling to stay profitable even in the Golden Era

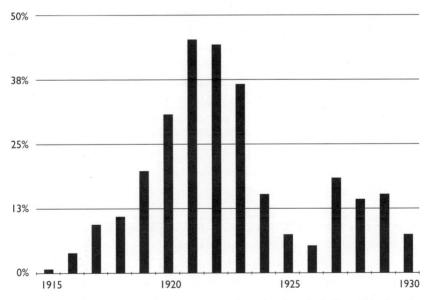

FIGURE 17. Estimated annual percentage cotton crop loss from the boll weevil in Georgia, 1915–1930. Data from Willard A. Dickerson et al., eds., *Boll Weevil Eradication in the United States through 1999*, Cotton Foundation Reference Book Series, No. 6 (Memphis: Cotton Foundation Publisher, 2001), 614–15.

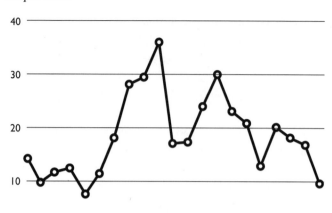

FIGURE 18. Georgia cotton prices, cents per pound, 1910–1930. Data from USDA Bureau of Agricultural Economics, "Statistics on Cotton and Related Data," Statistical Bulletin 99 (June 1951), 136.

had little means to make up for the added costs of the boll weevil. "Thus the golden hue of prosperity that appeared to be covering Georgia in the first decade of the new century," Willard Range wrote, "was not so real as it seemed to be, and the costs incurred fighting the boll weevil after 1918 made more severe an already bad situation."[6]

Though Range made this point in 1954, few historians seem to have picked up on his appraisal. If the boll weevil had never appeared, Georgians on the state's least productive land would have quit cotton growing anyway. They could not sustain cotton farming on any land that did not produce cotton at a high yield. Though many must have seen the writing on the wall, few admitted that cotton farming in all but the state's best land was dying. As the pest approached and then finally invaded, state leaders made the appearance of preparing Georgia's farmers to fight the weevil, though much of their work used the boll weevil's threat as a means to reach farmers about other issues, principally modernization and diversification.

*

The responsibility of educating farmers about the coming of the boll weevil fell to Georgia State College of Agriculture president Andrew M. Soule, who was happy to pick up the fight. Soule, a native of Canada, had worked in the extension services of Tennessee and Texas and as dean of the College of Agriculture at Virginia Polytechnic University before moving to Athens to become president of the Georgia College of Agriculture in 1907.[7]

Soule took over a few years prior to the boll weevil's arrival in the state, but he wasted no time in making the boll weevil battle the central part of the school's program. He recognized even before the pest's arrival that it held the key to increased funding and support for his school. His first act as president was the planning of a cotton school for January 1908. The college invited farmers from across Georgia to Athens for eleven days of faculty lectures and hands-on work at the school's research farm. The college required tuition of one dollar, so that, in Soule's words, it would be "virtually free to every farmer and farmer's boy in Georgia." The sessions were designed to teach Georgia's farmers three basic principles: maximizing cotton yields, diversification, and self-sufficiency. Soule himself delivered a series of lectures at the institute, including one titled simply "The Mexican Cotton Boll Weevil." He warned farmers against importation of weevils through seed and hull shipments from infested states, but he also linked general improvement of Georgia's farms to the coming of the pest. It was a strategy he would employ for several years.[8]

Following the first cotton school, Soule directed the college's efforts to take information directly to farmers via the state's rail system. An educational train began annual tours of the state, reaching rural towns with the college's message of modernizing farm operations and devoting less land to cotton. It too featured lectures on the boll weevil, designed to grab the attention of nervous farmers, but most of the information was on improving farmers' methods of raising stock and growing crops other than cotton for profit. The following year, in reaction to the positive response of farmers to the train, the state legislature appropriated $10,000 for extension education. In 1911, the body raised its appropriation to $40,000. The greatest boost for Georgia's extension system came with the 1914 passage of the Smith-Lever Act. The following year, Georgia received $35,174 from the federal government, the greatest of all of southeastern states, trailing only Texas and Illinois nationally. The State College of Agriculture immediately put these new funds to use.[9]

In addition to the college's work directly with farmers, Soule and his staff enlisted the support of the state's most powerful business interests. In November 1910, the College of Agriculture cosponsored with the USDA and the Atlanta Chamber of Commerce a meeting in the capital city. Planters, state and federal educators, and Georgia's most powerful businesspeople discussed what the weevil's eventual invasion of Georgia would mean for the economy of the state. Participants not only talked of methods of combating the weevil, but also of ways to assuage the fears of the state's cotton farmers. Attendees, especially those business leaders whose livelihood depended on an active cotton market, feared that landowners would abandon cotton completely when the boll weevil arrived, rather than learning to grow cotton under weevil conditions. Merchants, bankers, and agents of the state's railways recognized that the weevil could get the attention of the state's farmers, and, as a result, they too attempted to use the pest as a means to push their own agendas. Later, the Central of Georgia Railway distributed a bulletin up and down its line with the alarmist title ""The Boll Weevil is Coming! What Are You Going To Do About It?"[10]

While the school courted the help of the private sector, it increased its direct communication with farmers. The college continued to hold farmers' institutes in Athens, hoping to use the boll weevil, and farmers' own anxiety about the pest, to entice farmers to hear about alternatives to cotton. From 1910 to 1930, the State College of Agriculture sponsored nearly one hundred meetings that claimed to be concerned in some way with the cotton pest. In most cases, however, the main point of the meetings had little to do with the boll weevil. The cosponsors of these conferences revealed the true intentions behind the meetings. In 1917, for instance, the Georgia Dairy and Live Stock Association, the State Horticultural Society, and the Georgia Breeders' Association convened a conference ostensibly devoted to the cotton pest. The content of the meeting's message, however, had more to do with raising livestock, breeding animals, and growing alternative plants than it did with fighting the insect.[11]

In addition to the farmers' institutes and bulletins organized and published by the college, in 1912 Soule dispatched several agents to the infested areas of Louisiana, Mississippi, and Texas. These researchers studied not only how the weevil destroyed cotton and the methods used locally to limit the pest's effects, but also the broader social and economic changes brought by the weevil. Echoing the line repeated across the South about the weevil acting as a "blessing in disguise," agents reported that not all change brought by the weevil had been bad for rural society:

> The boll weevil in Louisiana and Mississippi has destroyed absentee land-lordism; he has helped to introduce livestock and a crop rotation system of farming; he has forced the farmers to grow their home supplies, which has largely done away with the credit system; he has brought to the large farm-

ers competent superintendents, and has awakened the small farmers of the country to the necessity of a better system of farming.

Intrigued by the agents' reports of diversification and modernization, in 1914, 1915, and 1916 Soule sent more farm educators and researchers to Alabama and Mississippi. For the 1915 journey, Soule enlisted 150 South Georgia farmers to travel with his agents to explore local conditions in the weevil-infected regions.[12]

Sending agents to the infested areas and preaching diversification at conferences was Soule's carrot. The widespread distribution of informational bulletins that scared readers into thinking that the pest would destroy, or had destroyed, all of the state's cotton was his stick. The college used informational literature prior to the entry of the boll weevil to the state, but as the pest moved closer, the school increased both the frequency and total numbers of bulletins printed. In 1916, the first year the boll weevil did measurable damage to Georgia's cotton, the college printed and distributed seven different circulars dealing with the weevil. Again, however, just as the local farm schools had used the boll weevil as a ploy to attract farmers to hear about a different farm topic, most of the state's bulletins dealt not with the pest directly, but used its arrival as a means to push for crop diversification and other farm improvements.

In November 1916, for instance, the college advertised its "short courses" on farming with a bulletin with a huge photograph of a boll weevil emblazoned on its front cover. Below the picture, a bright red caption (the first instance of color used in any bulletin) declared "Get Ready for the Boll Weevil. Make No Mistake." The pamphlet attempted to get readers' attention with the giant weevil, though its content had very little to do with the bug itself. The college published similarly misleading bulletins with catchy titles like "Starve the Boll Weevil," "Beating the Boll Weevil," and one that declared in boldface simply "POISON BOLL WEEVILS."[13]

Despite preparations, by the early 1920s the boll weevil was exacting a heavy toll on the state's crop. As Figure 17 illustrates, from 1920 through 1923, the weevil destroyed more than a quarter of the crop annually. By 1922, Soule admitted, "the advent of the boll weevil was a staggering blow to the farmers of Georgia." He added, however, that the weevil's damage was great because of the failure of the state's farmers' to heed his department's warnings:

In spite of all the efforts made to prepare them for its onslaught, the cry of "Wolf! Wolf!" so frequently heard and discounted, lulled them into a false sense of security. Like all plagues which have afflicted humanity, the weevil struck suddenly and with devastating fury.[14]

The pest had not, of course, struck suddenly. Georgians had known it was coming for nearly twenty years. No farmer in the state was surprised to find the pest damaging his or her crop in the early 1920s. In fact, if any entity had cried wolf, it was Soule's own department, which consistently used the coming of the weevil as a means to talk to farmers about issues having little to do with the pest. Of course, Soule's contention that the weevil had appeared overnight was another attempt to recontextualize the pest's arrival, to give the insect a kind of power that would in turn bolster his department's own authority. By playing up the pest's threat prior to its arrival, and by suggesting that it was an unstoppable force after the weevil was present in the state, Soule was using the pest as a scapegoat for whatever disintegration of the state's cotton economy was occurring.

Soule and the College of Agriculture were not alone in portraying the boll weevil as an invincible, devastating force. In May 1922, Harvie Jordan, a Georgia planter and board member of the American Cotton Association, wrote John Judson Brown, Georgia's agriculture commissioner, ostensibly asking for advice. The first two pages of the three-page letter gave the impression that Jordan's main reason for contacting Brown was to impress upon the commissioner that the boll weevil was exacting devastating damage on southern cotton farmers' livelihoods. Planters, Jordan told Brown, were in dire straits. The large landowners whom Jordan represented had become "insolvent" due to the "widespread invasion of the cotton boll weevil over the cotton states east of the Mississippi River." The beetle was no small threat; it "presents an economic problem in our agricultural industry which demands a complete change in the customs and habits of the farmers." It was a threat to the very foundation of the southern cotton economy: "The negro, the mule, an extensive acreage of cotton and the supply merchant are rapidly becoming a tradition," Jordan wrote. The weevil had, he argued somewhat contradictorily, fostered both a "heavy and continuous" exodus of tenant labor, and at the same time forced large landowners to move into towns, "leaving their lands at the thriftless mercy of ignorant tenants who are incompetent to cope with the boll weevil." Small farmers were "giving up in despair and leaving these lands idle and unproductive." Nothing less than a "complete change" in the way farms operated could save southern cotton production.[15]

The final page, however, revealed Jordan's true reasons for emphasizing the damaging presence of the weevil. Jordan and the ACA were not calling for Brown's aid in the boll weevil fight, but for his support of a new crop credit plan. Rising labor, pesticide, and fertilizer costs, along with falling market prices for the staple, had made cotton farming painfully expensive, even for the South's largest planters, Jordan argued. The ACA sought greater access to credit for the region's planters. Jordan also asked for Brown's support for a federally mandated minimum cotton price and his opposition to legislation limiting agricultural workers to an eight-hour workday. Jordan's rhetoric about the boll weevil was merely to attract Brown's attention by painting a dismal picture of southern agriculture that would spur action on mostly unrelated cotton marketing and labor issues.[16]

Powerful organizations of large landowners, like the ACA, were not the only groups using the weevil to press their own disparate agendas. Brown, in fact, used the pest to advance a cooperative farming plan. When a farmer from Union Point, Georgia, wrote Brown asking for his help in finding a more affordable solution to the boll weevil than applying the expensive poison calcium arsenate, Brown's answer had little to do with chemicals. "I *can* say to you, in all seriousness," Brown replied, "that we can grow cotton in Georgia in spite of the boll weevil *but* it cannot be done at the prices now prevailing, or anything like it, without losing not only our crops but our lands as well." Only if farmers were willing to sell cotton at "a price to cover cost of production plus a reasonable profit" would Brown recommend the crop. Those who sell their cotton at any price "insist that we take a gambler's chance at staying out of bankruptcy."[17]

The solution Brown offered this farmer and many like him was a cotton growers' cooperative. He again put the boll weevil at the heart of his rhetoric, though it was not actually a main facet of his plan. The pest had raised the cost of production beyond what the crop earned at market, Brown pointed out. Skimping on the application of pesticides was certainly not an option for farmers, but prices for the poison were constantly on the rise. "So you see we cannot grow under boll weevil conditions with Calcium Arsenate, and cannot grow without." He advised the farmer to join a cooperative and to demand that his neighbors do so as well. If farmers across the South refused to sell their cotton below a grower-determined price, and "say to the World: 'You cannot get our cotton unless you pay this price,'" then the price would automatically climb.[18]

The cooperative movement had had limited success in Georgia since the 1890s, when the Grange and Farmers' Alliance organized farmers to boycott using jute to wrap cotton bales, and helped farmers organize rural cooperatives throughout the state. Though these organizations gained hundreds of active members, most failed because of a combination of poor business practices and the refusal of powerful urban-based industrial interests like railroads and banks to give in to the cooperatives' demands.[19]

In his attempt to convince Georgia farmers to again join the cooperative movement, Brown added a powerful enticement to his argument: the final destruction of the boll weevil. If landowners followed through with their threat to cease cotton growing for an entire season, not only would prices soar, but "we would destroy the boll weevil absolutely," ridding the "American continent" of the insect. "And if it takes that to destroy the boll weevil," he added "and to teach the cotton consuming world" that farmers "cannot afford to raise cotton at a loss," then the weevil will be dead and prices will climb.[20]

The real dilemma for the state's cotton growers in Brown's view was the disjuncture between producers and marketers, not the boll weevil, but he continued to name the pest as the instigator of this trouble. In his 1922 state report, he pointed to the "continued ravages of the boll weevil," which make "permanent co-operative marketing associations" necessary. He admitted that the crop was "yet our money crop, and will be possibly for many years to come," despite the pest. "It is no longer a question as to whether Georgia can or cannot produce cotton under boll weevil conditions," he wrote. "It is an established fact . . . that we CAN grow cotton at the rate of one bale per acre on lands that were capable of producing one bale per acre before the advent of the boll weevil." In order to beat the weevil, however, Georgians would have to make major changes in the way they marketed the crop, not in they way that they planted it or protected it from insect ravages. "The great question now to be settled," he concluded, "is whether the cotton consuming world is willing to pay cost plus a reasonable profit for cotton, or whether they will insist that the farmer continue to grow cotton for less than cost of production."[21]

Brown's ideology, however, did not match his own farming record. All the while recommending that cotton farmers band together to sell their crop, or that they cease cotton production altogether because of low prices, Brown directed his brother, who managed Brown's personal plantation, to plant more and more cotton. In 1922, he wrote J. Polk Brown, demanding his brother "PUT THE CALCIUM ARSENATE TO THE COTTON." Prices for the staple were climbing, this supposed proponent of diversification told his brother, and he wanted to have a bumper crop and sell it while prices were near their peak. "Cotton went up awfully today and the world has got a cotton famine right on," he wrote, adding that if they sold the crop for a good price, "I hope to eat fried chicken and have some good old milk and butter and really enjoy myself."[22]

Following a visit to his farm, however, Brown was apparently less than impressed with his brother's management. "Glad you are well and getting along all right," he wrote upon his return. "But, I am afraid you are not dusting the Cotton like I want it done. There is absolutely no necessity of allowing the boll weevil to ruin a single little boll of cotton." Brown advised his brother to tell the neighboring landowner to "put the dusting on his and to do it NOW" or else the "b.w. will back-work the cotton and ruin it." "NOW is the time to fight like the house was on fire," he concluded.[23]

Even for the state's leaders in the modernization and diversification effort, the pull of periodic spikes in the cotton price was too enticing to resist. Brown was willing to ignore his own advice in favor of the prospect of one more bumper cotton crop. As the weevil's damage increased through the 1920s, farmers, educators, and entrepreneurs sought ways not only to cash in on one more cotton harvest; many sought ways to take personal economic advantage of cotton's demise. Farming was the leading industry in the state and there was no shortage of people who saw the arrival of the boll weevil as a means to profit from the changes it was bringing. If cotton was truly dying as rural Georgia's primary cash crop, what would replace it? How would the alternatives be grown and marketed?

James William Firor was one educator-turned-entrepreneur who thought he had the answer. Firor was in many ways a typical twentieth-century farm educator. He formally studied agriculture at the Maryland Agricultural College and upon graduation in 1908 found a job with the West Virginia extension service. He moved to Athens, several years later, to join his brother on the staff at the Georgia State College of Agriculture. Shortly after the breakout of war in Europe, Firor volunteered for the army, and served in France for nearly two years. Upon his return, he rejoined the faculty and settled into his job as an extension agent.[24]

As a farm researcher and teacher with the responsibility of telling landowners how to prepare for and fight the boll weevil, Firor found himself constantly away from his fiancée, traveling through the state visiting farms. He grew frustrated not only by the constant travel, but by farmers' unwillingness to consider alternative crops to cotton. When Robert Schmidt of Yorktown, Virginia, wrote Firor asking about the prospects of employment as an extension agent in Georgia, Firor replied with uncharacteristic bluntness:

I have just returned from a trip through the southern part of Georgia and feel somewhat weak in the summach [*sic*]. You have heard of an insect called Boll Weevil. What he did to the cotton in the southern part of Georgia is hardly possible to relate in a letter under existing postal laws and regulations . . . The season was very favorable to the weevil and unfavorable to cotton.

Firor advised Schmidt to steer clear of the cotton South entirely.[25]

The agent's frustration had been building from the moment he returned from France and pushed him to begin his own search for a new job. He applied for several similar extension jobs outside the boll-weevil-plagued South, at schools in Illinois, Florida, and Kansas, and also inquired about jobs in the private sector. By the end of 1919, Firor was a newlywed looking to build a family, but trapped in an exhausting job with little reward.[26]

Firor knew that Georgia cotton farmers, especially those on marginal lands, were standing on their last leg. Recognizing that these growers could not make the change from cotton to a different cash crop without substantial aid from seed suppliers, marketers, and buyers, Firor began looking for another career that might take advantage of farmers' exodus from cotton. In a letter to Atlanta businessman M. C. Gay, Firor laid the groundwork for a project that he thought would bring financial success and stability. Firor attempted to convince Gay that sweet potatoes were in the process of replacing cotton as the region's important cash crop. Putting the boll weevil's devastation at the center of his pitch, Firor wrote to Gay that "the production of sweet potatoes is at the present time being greatly stimulated by high prices and by the fact that farmers throughout the cotton belt are looking to [alternative] crops to take up some of the acreage which was formerly planted to cotton." Firor estimated that if farmers had the support of a company to supply the proper seed, to store the sweet potatoes after harvest, and to take the product to market, then the tubers would eventually replace cotton as the king of southern agriculture. He pressed Gay to help him form a business that would guide farmers through the transition to sweet potatoes.[27]

Less than a year later, Firor resigned from the College of Agriculture and became manager of the Planter's Products Company, based in Montezuma, Georgia. A company statement made the boll weevil the cornerstone of its marketing:

The experience of farmers during the last few years, under boll weevil infestation, has shown that cotton alone is not significantly dependable as a money crop for south Georgia, and most of us who have grown cotton under the boll weevil conditions, will doubtless agree that your community will not remain prosperous unless we diversify our farming program so as to have at least another money crop. The question therefore arises—what other crop shall we grow?

To no reader's surprise, the answer Planter's Products offered was sweet potatoes, and the company was available to provide all of the apparatus necessary to grow, store, and sell the tubers.[28]

In private letters, Firor explained the economic motivation behind this sweet potato "opportunity." He knew firsthand that Georgia's farmers were skeptical of diversification talk—if cotton was to be replaced, it could not be done with simply a variety of fruits and vegetables. Farmers sought a single cash crop to replace cotton. As a result, Firor emphasized the economic benefits available to the sweet potato farmer. He thought that once farmers believed in the potatoes, he could easily fill his company's new warehouses. Once his company was supplying seed, know-how, marketing help, and storage to these former cotton growers, Firor expected business to take off.[29]

Despite his hopes for independent success, five years after he left the College of Agriculture, Firor was back on its staff. For reasons unexplained in his correspondence and personal papers, Planter's Products failed. There could be dozens of reasons why the company did not succeed, but at the very least the company's demise suggests the difficulty that Firor and his partners faced in convincing south Georgia cotton farmers to grow and market sweet potatoes. Beyond the social and cultural significance of cotton to rural Georgia, a switch to sweet potatoes would have necessitated an upset of the rural economy itself. Firor had attempted to capitalize on the destruction caused by the boll weevil, but in the end, most farmers were still committed to growing the staple that the bug attacked. Even those who wanted to diversify may have lost their access to credit after a single devastating cotton season. Acres and acres of Georgia farmland turned to timber in the late 1920s, too eroded to farm and lacking laborers to work it. By the end of 1925, Soule found himself back in the fields representing the state's College of Agriculture, talking to growers about the best way to beat the boll weevil.

Years before Firor's failed experiment, Andrew Soule had addressed a group of farmers and asked, "Can we whip Billy Boll Weevil?" "Undoubtedly we can!," he had replied. Fifteen years later, the insect had destroyed thousands of tons of Georgia cotton. Diversification efforts had stalled. Reports of a vast outmigration of rural workers had gripped local and national headlines for at least a decade. Yet still Soule was ready to publicly declare victory over the weevil. In the spring of 1930, in the *Atlanta Journal*, he asserted that after a decade of agricultural depression caused by the boll weevil and low cotton prices, "the days of pessimism are behind us." Soule referred to the success-ful cotton crop of 1929, and foresaw a future of cotton growing in the South similar to its now relatively distant past.[30]

*

In one sense, the agricultural educator's predictions of cotton's resurgence proved right. Cotton production had by 1930 rebounded to the levels of the 1890s and early 1900s, but there was one major difference in rural life. Since the earliest days of the pest's arrival in the state, Georgia's farm experts, like those in states across the weevil territory, had been very concerned about a labor exodus, and in Georgia, the outmigration had been heavy. As Soule penned his boll weevil victory speech for the Atlanta paper, the stream of African American migration from the rural areas of the state continued. For Soule, however, this was not a tragic turn of events spawned by the weevil.

Much of Georgia agricultural officials' fear about labor exodus was based upon their own long-term reputation for creating policies damaging to the plight of landless farm workers. Agriculture officials had not ignored tenants for generations, but instead crafted a vicious system plainly intended to limit black landownership opportunities and to tie African Americans to the rural areas. Thomas P. Janes, Georgia's Commissioner of Agriculture, set the tone for the state's policy towards black laborers as early as 1875 when he com-mented on the prospects of black landownership:

> It is not reasonable to suppose that men, naturally indolent, ignorant and superstitious, mere muscular automata by habit, having been accustomed to direction even in the minutia of their work, could, by a presidential proc-lamation, be converted into intelligent and reliable business managers.

The spirit of Janes's remarks guided the state's treatment of black tenant farm-ers well into the twentieth century. The extension service's treatment of the state's tenant population exacerbated the already precarious ties binding farm-ers and tenants. For small landowners, there was little help that the state could provide to ensure that they or their tenants survived the weevil's onslaught. For tenants, however, changes in the fate of small landowners who they worked for could be more fickle and devastating than those who worked for large planters.[31]

Minnie Stonestreet of Washington, Georgia, was not a typical farmer. In 1924 she was an educated, unmarried woman, when to her surprise, she inherited a 190-acre farm from an uncle. Stonestreet jumped at the chance to be an independent farmer. "How proud I was over owning a farm," she later wrote, "a plantation all my very own." With "dreams of a fortune made farming," Stonestreet "set about to make those dreams come true." After selling timber from the land for six hundred dollars, she negotiated with Lee Slakey, "an old Negro man and his wife whom he called 'Pig,'" to rent part of the land and to help put in the crop.[32]

The three farmers went to work on the land, planting cottonseed in early spring, and chopping the weeds that grew as summer began. For Stonestreet, farming meant a kind of personal independence she had always desired. "Why the first time my tenant came driving my mules to my wagon, I felt like a millionaire!" she recalled. Part of this feeling came with hiring tenants. With cash from a sale of timber from the land, Stonestreet had enough money to pay her tenants at the beginning of the month "their rations for 30 days." Despite feeling rich, Stonestreet ran out of cash by midsummer, and like most landowners began to borrow against her cotton crop. Spring planting had gone well, but "then came the summer." A small number of boll weevils appeared early in the season, and they multiplied into an enormous throng by late summer. At year's end, Stonestreet recalled, "Lee Slakey, the negro farmer, came to the office with the gin certificates for all the cotton grown on my place that year." Instead of realizing her "dream of a fortune made on a farm," the tenant reported to her that they had ginned very few bales. Stonestreet had "nightmares of acres and acres of cotton with all the people I owed standing in the middle of them."[33]

Having gambled against the boll weevil and lost, Stonestreet saw her life quickly spiral out of control. With no cash, she stopped payment on her house insurance, only a few weeks before it caught fire and burned to the ground. Over the winter of 1925, one of her two mules died along with most of her chickens and hogs. Though the effect of these tragedies on Stonestreet was damaging, for Lee and "Pig" Slakey, it closed all hope of staying on the land for another year. Stonestreet could move home to live with her mother in town, but the tenant farmers had to hit the road and search for a new place to live and work. Slakey, however, never made it off Stonestreet's farm; he died over the winter. "Pig" left on her own. Without tenants, Stonestreet knew she could not possibly recover her losses the following year. "With this last blow," Stonestreet later wrote, "like the drinking man who was several times thrown out of a party he had gone to uninvited, I picked myself up with the conclusion that fate did not want me to farm." She returned to her mother's house in town and eventually found government work with the Works Progress Administration (WPA).[34]

Though Stonestreet judged her farming stint a "grand failure," she did not find herself on the road in search of a new home and a place to farm, as so many tenants hurt by the boll weevil did. Janie Young, a middle-aged black farmer from Blythewood, South Carolina, would not be so lucky. Following an elaborate wedding celebration, complete with several cakes, "a white dress and a long white net veil," and "a big bouquet of white flowers," Young and her husband drove off from the reception under a shower of rice, bound for a new job on a nearby plantation.[35]

A landowner had hired Young's husband Nick to work his cotton, and Janie was to work in the planter's house. The landowner paid the newlyweds fifteen dollars per month for their labor and provided, in Janie's words, "a ramshackle old house to live in." From the time they arrived, the couple constantly sought ways to make (and retain) more money, to climb the mythical agricultural labor ladder. Janie found she could make more money working in the fields alongside Nick than in the house, so she did. "We could live very well with me working all de time in de field for forty cents a day," Janie later told a WPA interviewer, "I did anything dere was to do on a farm, 'cept plow. I sow de seed, chop cotton, hoe de crop, and put down fertilizer, and do anything else dey wants done."[36]

The Youngs scraped by as wage laborers on a section of Mr. Wilson's land for four years, scrimping and saving every penny they earned until they had saved enough money to buy a mule and a wagon. With these tools, the Youngs bargained that they could make more money renting a piece of land than laboring for a set wage. For a poor couple with a roof over their heads and a small but steady income, this decision was no small matter. It was, however, a choice that thousands of southern farm laborers who hoped to climb out of wage or sharecropping labor arrangements made each year. Moving was always risky, but tenants looked for better jobs after both bad times and good. The Youngs' gamble, however, could not have been made at a worse time.[37]

The couple found a landowner, Mr. Wall, who agreed to rent fifteen acres to the Youngs for a one-time payment of five-hundred pounds of lint cotton due at the end of the season. With a mule and tools, the Youngs could limit the amount of debt they took on at the beginning of the year, and could make back what they borrowed and owed in rent at the end of the season. The arrangement had major advantages over their previous deal. The couple acquired limited debt for the use of a plow, cottonseed, and other necessities for raising a crop, and they tilled and planted the land. Unfortunately for the Youngs, boll weevils made their first appearance in the area late in the season. Janie remembered "dat was de first year de weevil was so bad, and we didn't make no cotton to speak of. We didn't have near enough to pay de rent." The tools and savings that the Youngs had spent five years accumulating during their climb from wage laborers to renters vanished with one weevil-heavy season. In debt to Mr. Wall, the Youngs agreed to stay and try work as share-croppers the following year in an attempt to square the balance, slipping back down the mythical ladder.[38]

The Youngs' example demonstrates how difficult it was for sharecroppers merely to become renters, let alone landowners. For African American ten-ants, there was no shortage of obstacles to landownership. Though there is a good deal of scholarship that explains these restraints on ownership in great detail, the obstacles basically came down to a wicked combination of social, economic, and political factors. Banks often refused credit to African Ameri-cans to buy land because many white bankers simply did not believe that African Americans could profitably farm their own land. Others might refuse loans to African Americans because landownership was an expression of eco-nomic power that some whites found threatening. As one sharecropper put it, "It ain't what I owes, it's gettin' to owe." Politically, those African Ameri-cans who actually bought land had few ways to ensure its protection. Though some of the state's white leaders, including agricultural commissioner Brown, suggested that the state should encourage black landownership with progres-sive legislation, these voices for land reform were rare and ineffective.[39]

Sociologist Arthur Raper found that those African Americans in Greene County, Georgia, who became landowners usually did so only thanks to the direct aid of some prominent white landowner. Some 75 percent of black owners bought their land from former landlords, who in most cases initiated the sale of land. "In some instances the white landlord liked a particular Ne-gro and helped him become an owner," the sociologist explained, and "in other cases they needed because of debts to sell off part of their land." "Only in rare instances," Raper argued, did black landowners purchase a farm "on the open market." From the turn of the century to 1920, as Figure 19 shows, as cotton production increased in the state and the boll weevil was at most a minor threat to Georgia cotton, black landownership increased slightly, though tenancy levels increased markedly.

As discussed in the opening of this chapter, the increases in cotton pro-duction from 1900 to 1920 did not bring with them riches, and by the time the boll weevil began destroying large portions of the crop in the early 1920s,

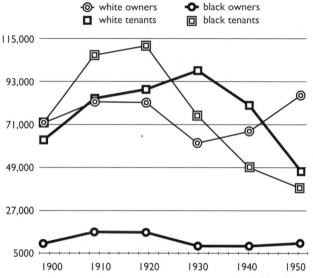

FIGURE 19. Land ownership and tenancy by race, Georgia, 1900–1950. Data from Neil Fligstein, *Going North: Migration of Blacks and Whites from the South, 1900–1950* (New York: Academic Press, 1981), 84–85.

tenants' slim options for landownership were disappearing. With the open market closed to African Americans in search of their own land, plus the increased burden of farming in the presence of the boll weevil, the oppressive credit system, Jim Crow segregation, and racial violence, many black Georgians decided to quit farming altogether. The timing of the resulting outmigration, which coincided with the weevil's worst years, became the subject of a number of scholarly investigations that publicized the notion that the weevil alone pushed millions of black Georgians from the land.

For instance, the most revealing trend in Figure 19 is not the slight rise in both tenancy and ownership levels from 1900 to 1920, but the plunge in the number of African American tenants in the state following 1920. In fact, Georgia's entire population grew only 0.4 percent from 1920 to 1930. (Montana was the only state with a lower percentage of growth.) It was this drastic change in rural life—the exodus of nearly two hundred thousand black Georgians from rural areas over that decade, which coincided with the worst years of the boll weevil invasion—that would foster the notion that the insect was to blame for the rural upheaval.[40]

Indeed, the migration was especially important to Raper's Greene County study. Raper was but one of a cadre of prominent social scientists who studied rural Georgia during the early twentieth century whose findings perpetuated this idea. Howard Odum, Will W. Alexander, Thomas J. Woofter, Charles S. Johnson, and Raper worked both together and independently on a number of studies of life in the Georgia countryside from roughly World War I to World War II. Arthur Raper's *Preface to Peasantry: A Tale of Two Black Belt Counties*, the best known and most important of these books, concluded that the boll weevil was a principal factor not only in the demise of cotton culture in Black Belt Georgia, but in the exodus of African Americans from rural areas as well. It was a conclusion endorsed by the bulk of subsequent scholarship, which has relied heavily on Raper's research.

Indeed, the statistics surrounding Raper's research are dramatic. He devotes an entire chapter of *Preface to Peasantry* to the "exodus" of African Americans from Black Belt Georgia. (Note that Raper uses "Black Belt" to refer the area of the South with the highest African American populations, not to the Black Belt soils discussed above. The latter are limited mostly to Alabama, but are found in small sections of Mississippi and Georgia, though not in the part of northern Georgia under study by Raper.) In it, he argues that the "immediate causes of the exodus" were the county's overdependence on cotton and subsequent boll weevil devastation. "The winged demon," Raper writes, "had descended upon the planters over night." In fact, from 1917 to 1919, Greene County experienced a boom in cotton production. As prices in the state peaked, its farmers had near-record harvests reaching twenty-thousand bales in 1918 and 1919. In 1920, the boll weevil arrived, but Greene's farmers continued with cotton. By the end of the season, the insect had devoured a third of the previous year's total. By 1921, the county's ginned cotton level reached only 1,487 bales, down from twenty-thousand two years prior. In 1922, farmers managed only 333 bales countywide. In Raper's rendition, "Debts and taxes went unpaid; credit vanished; chaos reigned."[41]

Considering these colossal crop losses, it is no surprise that over the decade of the 1920s, as Raper notes, a quarter of Greene County whites and 41 percent of its African Americans exited. Though the migration of black farmers from Greene County was major and important, Raper himself admits that it was far from typical. In Macon County, far fewer tenants left. Greene farmers had been unlucky, Raper argues, because the weevil's three most vicious years were consecutive. "One lean year consumed the fat of the previous years," he argued, and three bad years together could rock the economy of the whole community. In the county seat at Greensboro, two major banks failed during the weevil's worst years. In Macon, however, none of the weevil's worst years fell back to back, providing at least one season for farmers to make back their losses, or to diversify into a less risky crop. In fact, it was rare for any county across the South to experience consecutive years of devastating boll weevil damage. Greene County made for an illustrative, even dramatic example, but it was atypical—a fact few of the scholars who have relied on *Preface to Peasantry* have admitted.[42]

A sampling of other Georgia counties reveals that the outmigration of African Americans from rural areas during the boll weevil's most devastating decade varied significantly. Greene County lost 41 percent of its black population during the 1920s, but other heavy cotton-producing counties lost far fewer people. Macon County, for instance, which had roughly the same acreage of cotton as Greene County in 1920, lost only 7 percent of its black population over the decade. As one of Georgia's heaviest cotton-producing counties, Sumter had one of the state's largest African American populations in 1920, but its black population fell less than 10 percent over the decade. These examples demonstrate that Greene County's experience with the boll weevil and population loss was unusual, but the larger and more important point is that scholars' heavy use of Raper's example has distorted our understanding of the boll weevil's effect on outmigration.[43]

Though much has been made in United States historiography of the "Great Migration," this mass movement of rural southerners is still a misunderstood and understudied subject. As suggested in chapter 4, the migration of rural southerners had been a constant condition since the Civil War, and it was not limited to African Americans. As Numan V. Bartley wrote, while black Georgians moved in search of better lives, "[white] farmers, who lived in another man's house and worked another man's land, moved into the mill villages to live in another man's house and work another man's machine." Beginning immediately after the Civil War, because of constantly changing land, credit, and crop conditions, landless laborers had constantly been on the road in search of new or better work arrangements. The boll weevil had perpetuated this movement of laborers all along the frontier of its moving mass. In fact, as the boll weevil pushed slowly to the east from Texas towards the coast, labor moved against it. Cotton farming expanded in the western Cotton Belt, in places like Texas and Oklahoma where the soil was less worn out, and would still produce cotton, and many tenants from 1895 to 1920 had moved into these expanding lands.[44]

African American migration from the South to northern cities has its own long and complex history related to the spread of the boll weevil and its myths. Though this migration increased markedly during the First World War, it was an extension of a pattern of movement north that had been in place for generations. The pull of industrial jobs was one factor in migrants' decisions, but most individuals also felt a sufficient number of "push" forces that motivated them to leave. Many black tenants responded to the destruction of the boll weevil by packing up their possessions and moving, but it was rarely the boll weevil alone that made up one's mind. Steven Hahn has argued that despite most historians' understanding of the Great Migration as "the product of a very specific set of circumstances that coalesced in the mid-1910s," including the boll weevil, it in fact began much earlier. "We must remember," Hahn argues, "that a northward shift in black migration was already in evidence in the 1890s, and that it was closely connected to a substantial trend, beginning in the 1880s, that took growing numbers — sometimes temporarily, sometimes permanently — from the rural districts to the towns and cities of the South." The migration was not a single movement at all, but simply a coalescing series of individual decisions that men and women made about how their personal fates were linked to the place they lived. As Amiri Baraka has written, "It was a decision Negroes made to leave the South, not an historical imperative."[45]

As the number of African Americans leaving the rural South increased, white landowners who had been dependent on their work became concerned about a sufficient supply of labor. In the 1930s, Posey Oliver Davis studied outmigration from Alabama and found that among southern whites "the chief concern is not about where Negroes are going but (1) Why are they leaving and (2) What will be the effect of their going on southern agriculture?" Davis cited a plethora of reasons African Americans had for moving: "poor schools, extortionate charges of creditors, swindling, wretched homes, unfair suffrage laws, cheating in the handling of cotton, injustice of courts, boll weevils, and high wages elsewhere." In sum, however, this litany of factors was not reason enough. Davis concluded that "the main reason is an economic one," lower wages on cotton fields than in northern factories. Many white southerners embraced this explanation. The idea that the pull of northern industry was the only force at work in migration was a half-truth held particularly strongly by white southerners like Davis. Scholars, farmers, and politicians claimed that poor schools and housing were merely "minor" reasons for the outmigration. "This is evidenced by the splendid feeling existing between the Negro and White races in the South" Davis concluded. "Not since the Civil War has the feeling been better" between rural whites and African Americans.[46]

If, in addition to the pull of northern jobs, scholars allowed for any single southern factor to explain rural black outmigration, the boll weevil was the most convenient. "It is known that the migration of Negroes from the South has paralleled the spread of boll weevils," Davis pointed out. "Much destruction frequently followed the spread of boll weevils, and in many instances this has been followed by Negro migration." "Being well suited to cotton farming, they [African Americans] naturally become frantic when weevils make cotton-growing hazardous," the scholar continued. "Instead of trying to adjust their farming to meet boll-weevil conditions, they turn to industrial life." It was for Davis and others the "panic" of African Americans in the face of the insect that created this exodus. The bug, rather than poor schools, economic exploitation, Jim Crow social restrictions, and racial violence, was the real culprit. In other words, white Southerners had done little to foster the movement, could do nothing to stop it, and should not be too concerned about it. Davis concluded that black migration would end up having only a positive effect on the South.[47]

Other white southerners came to a different conclusion and worked to stem the outmigration. John Egerton concluded that at first, "white Southerners couldn't decide whether the exodus was a good sign or a bad one." Agriculture commissioner John Judson Brown, however, was sure that black labor leaving the farm was a bad sign for white landowners. As a member of the Ku Klux Klan, Brown urged his fellow Klansmen to do what they could to keep African Americans in the South. Many white Georgians obeyed. Incidences of violence at rural train stations, along roads north, and within the black communities increased along with migration.[48]

Though these cases indicate that there were plenty of push factors in black migration from the weevil-infested South, academic scholars have been drawn to the pest as the major contributor. In addition to the work of Robert Higgs discussed in chapter 2, Warren C. Whatley and Gavin Wright's study of southern black labor in this period addresses the 1920–1923 "boll weevil depression" from the perspective of northern industry. Whatley and Wright found a surprisingly high level of African Americans hired during 1922 and 1923 in three different northern factories (Ford in Detroit, Byers Steel in Pittsburgh, and Pullman in Chicago). The factories hired very few African Americans prior to these years and a smaller number after, until the late 1930s. Though these companies were in a different industries, regions, and "different points in their life cycle," they experienced an increase in the *available* black labor during that year. The authors conclude that the explanation for this "striking example" was "on the supply side." The boll weevil had created such havoc by destroying Georgia cotton in this two-year period, they argued, that there was a dramatic shortage of ripe bolls to pick in the fall, meaning tenants and part-time pickers were not needed. Thousands of Georgians boarded trains for the North, where employers reacted to this brief interruption in the labor supply by hiring the southerners. Interestingly, these factories did not again hire massive amounts of southern black labor until the late 1930s and 1940s.[49]

Studies like Wright and Whatley's that trace specific points of departure and destination at precise time periods are rare. Unfortunately, most scholars have relied on decennial census figures and bolstered their arguments with anecdotal evidence. A number of these scholars turn to Raper's Greene County findings to provide a local perspective because of convenience. Historians of the early twentieth century South have indeed relied too heavily on Raper's study, and readers of the literature on Georgia must surely be growing tired of hearing about Greene County almost to the exclusion of the rest of the state (and region).

*

The misconception of the weevil's impact on migration is not based on scholarly work alone. Songs about the boll weevil grew in popularity both around the South and across the nation by the 1920s, and their main theme remained movement. In almost every version the beetle was "lookin' for a home." In her study of migration narratives, Farah Jasmine Griffin admits that the artists who created these stories—whether visual, musical, or literary—cite explanations for why and how they moved from the South that are rarely accurate. Nevertheless, to the audiences that encountered the song in Atlanta nightclubs, rural honkytonks, or on radio and later television, the accuracy of a singer's boll weevil story meant little.[50]

Though farmers and cowboys in Texas first performed these songs as early as the 1890s, by the mid-1920s people around the country could hear and even buy the boll weevil song. Professional recordings, like the first version sung by Mississippian Charley Patton, appeared in "race record" stores throughout the country beginning in the 1920s. By 1930, the song could be heard in the repertoire of dozens of prominent blues and folk singers, both in concerts around the country and on recordings.

Despite the popularity, the message about migration remained the same. In Texan Huddie "Leadbelly" Ledbetter's version mentioned in chapter 2, the weevil survives a bevy of attempts on his life, and the frustrated farmer refuses to sell his lone remaining cotton bale to the merchant. If he retains the bale, "I'll have a home," the farmer promises himself, "I'll have a home." The farmer then ends the song by "moving on."

> If anybody asks you people who sang you this song
> Tell 'em it was Huddie Ledbetter,
> He done been here and gone.
> He's lookin' for a home,
> He's lookin' for a home.

This theme of migration and constant struggle for a new place to live and work permeates nearly all versions of the song from the first-half of the twentieth century, and it left an indelible impression on many about the history of the boll weevil.[51]

Though Patton's recording had a sound and substance that made it distinctly Deltan, the first wave of weevil records appeared when Georgia, not the Magnolia State, was ground zero for the pest. In fact, Georgia's two most influential early blues performers, Blind Willie McTell and Kokomo Arnold, recorded versions of the boll weevil song in the 1920s as farmers in their home state battled the bug. Arnold's "Bo-Weavil Blues" became widely distributed in the 1930s. Though Arnold follows the standard tune and form of the traditional boll weevil song that had been in existence for decades, his lyrics have very little to do with country life. The song reflects the growing migration of rural Georgians to urban areas. In fact, the boll weevil itself has lost its affection for cotton in the song:

> Now Mister Weevil, how come your bill's so long?
> Now Mister Weevil, how come your bill's so long?
> Done eat up all my cotton, started on my youngest corn.

There is more evidence of urban life in Arnold's version. Not only could the pest put pricey health care remedies (or recreational drugs) out of reach, the mill life that so many former cotton laborers found themselves in by the mid-1920s itself had become another victim of the weevil.

> Says the merchant to the doctor, "Don't sell no mo' C.C. Pills"
> Says the merchant to the doctor, "Don't sell no mo' C.C. Pills"
> 'Cause the boll-weevil down here in Georgia done stopped all these cotton mills."

Arnold's narrator seeks a remedy by pleading with the weevil to tell the world what it has done to the Georgia farmer:

> Now Mister Boll-Weevil, if you can talk why don't you tell?
> Now Mister Boll-Weevil, if you can talk why don't you tell?
> Say, you got poor Kokomo down here in Georgia catchin' a lot of hell.[52]

Like Arnold, Blind Willie McTell played his version of the song mostly for urban audiences in Atlanta, crowds that undoubtedly contained many migrants from around the rural South. McTell's song tells a history of the pest's movement, linking it to the migration of those who worked cotton fields and now found themselves listening to the singer on the streets and in the clubs of Atlanta. As the song begins, the speaker asks the boll weevil, "where you say you get your great long bill?" The pest replies

> I got it from Texas, out in the western hills.
> Way out in the panhandle, way out in the western hills.

As the song progresses, the weevil comes to represent more than simply a threat to cotton, but a potential destroyer of rural life. "Boll weevil, he told the farmer, don't buy no Ford machine," McTell sings, "you ain't gonna make enough money to even buy gasoline."[53]

Though the versions sung by these Georgians, McTell and Arnold, gained relatively wide audiences, no version from the state better links the boll weevil to tenant migration than that of Buster "Bus" Ezell, who performed publicly in the state into the 1950s. Ezell begins the song with a description of the weevil's ability to quickly multiply:

> Well the first time I saw a boll weevil he was setting on a cotton square,
> Next time I saw Mr. Weevil he had his whole family there.
> What you reckon he said?
> 'Bout to kill me dead.

By the third verse, the speaker is already destroyed, and is in search of a better place to farm:

> Well, I'm going back to Texas where I was bred and born,
> I ain't fond to leave Georgia, but Georgia ain't none of my home,
> I'm on my way, I'm on my way.

The speaker's allusion to a previous migration, from Texas to Georgia, underscores the gravity of his decision to move all the way back west. When the weevil later tells the frustrated migrant farmer, "Don't you lose your mind, Don't lose your mind," it is easy for the listener to understand the emotional wear that constant movement must have exacted on the tenant farmer.[54]

Though by the 1920s the song had found an audience throughout the South and parts of the North, and lost some of its rural context along the way, the decade that followed took the song even farther from its element. Thanks in part to Alan and John Lomax, who "discovered" folk music across the South during the 1930s and after, the boll weevil song became popular among singers who never set foot in a cotton field. Though the Lomaxes recorded dozens of versions of the song in southern prisons, work camps, and farmhouses, it was the generation of artists who heard these recordings and in turn took the songs to a new audience who are most responsible for spreading the boll weevil song and the myth it contained. Woody Guthrie, Cisco Houston, and Carl Sandburg began singing the song in the 1930s and 1940s. Though these leftist folk singers originally embraced the song's political message, they slowly turned its melody and content into a children's song. The choruses became more repetitive and humorous. It seems that the longer southern farmers were accustomed to the weevil's presence, the traditional song's narrative of crop destruction fell away, in favor of catchy verses about a seemingly silly bug.[55]

To singers and audiences alike, the idea that the boll weevil wiped out their cotton livelihood comes closer to the truth in Georgia than in any of the other Deep South states. The appearance of the pest did correspond with both a decline in the state's cotton production and an exodus of cotton laborers. At the local level, however, the weevil made its greatest impact on cotton societies already in decline before the pest arrived. Those locales with soil exhausted by a century of cotton growing, and by an increasingly vocal and powerful rural working class, found it hard to maintain cotton's supremacy. Farmers on much of the state's marginal farmland turned to other crops or raised livestock and chickens, and much of the land reverted to timber. The federal government even stepped into parts of the state to create national forests on land that had previously grown cotton. These were important, telling changes in the state's agricultural, economic, and social histories, but instead of pointing to generations of tenancy and land mismanagement, debilitating credit arrangements, and persistent racism, most southerners, white and black, chose to proclaim the boll weevil as the instigator of this great change. It has proven a powerful legend in Georgia and across the Deep South.[56]

The Boll Weevil's Lost Revolution

In July 2003 a story aired on National Public Radio's news program *Morning Edition* that heralded the final demise of the boll weevil in the United States. "A century-long war between American cotton farmers and one of their most dreaded foes, the boll weevil, is coming to a close," announced host Bob Edwards, and "it looks like the farmers are going to win." Reporter Dan Charles explained the history of the pest's move across the South, recalling that in the twentieth century, "cotton was the economic foundation of southern society" until the pest emerged and destroyed the region's entire way of life. "*Wherever* the boll weevil appeared," Charles explained, "landowners went bankrupt and sharecroppers abandoned their homes."[1]

The insect devastated the South and cotton became only a marginal crop, Charles said, until the 1980s, when scientists in Mississippi isolated the chemical that male boll weevils produce to attract females. Replicating this scent allowed researchers to place traps, baited with the scent, throughout fields to monitor local infestations. Farmers paired this monitoring system with heavy aerial applications of Malathion, a new breed of insecticide that had proven highly successful against weevils. Where and when the traps caught the insects, farmers covered the area with pesticide. Using crop dusters—planes that technologically bore little resemblance to those first tested at DPLC in the 1920s—farmers sprayed the weevils with this brave new pesticide. The airplanes were equipped with global positioning systems (GPS) to ensure that the poison made it to the exact places where the traps had indicated a weevil presence. Computers then matched the course of each airplane's flight to field maps to ensure that every single weevil was killed. The elaborate technique was remarkably effective and researchers sought to replicate it across the Cotton Belt. In 1983, with the help of the federal government and state extension services, but largely paid for by individual cotton growers, farmers began this monitoring and spraying system in Virginia and North Carolina, in an attempt to slowly push the weevil back to Mexico. By the summer of 2003, scientists could claim that the weevil had been eradicated from the entire Cotton Belt except Arkansas and Texas.[2]

The story's tone matched the optimism of the scientists interviewed, but almost a decade after it aired it seems almost foolhardy. Despite a sound byte offered by environmental historian Edmund Russell, who said that "we can never declare victory" in wars against insect pests, the spirit of the report was a celebration of victory. Eradication's boosters even promised the entire South would be free of the weevil by 2010.[3]

They were wrong. Bright green weevil traps still line all cotton fields in Mississippi and other Deep South states, as mandated by law. And the traps still collect specimens. (In 2009, USDA officials claimed that 2 percent of the nation's nine million acres of cotton still had boll weevils present.) But even as cotton acreage has declined along with its insect enemy, the boll weevil remains central to rural southerners' understandings of their modern identity and the history that produced it. As the preceding pages explain, the boll weevil's arrival in the South did not mean the automatic and complete collapse of the southern economy, despite the lasting power of the notion that it did just that. The myth of the weevil's destructiveness was born in the fields of Texas when the first few bugs began damaging American cotton, and it has proven to be powerful and enduring, as the NPR report shows.[4]

One reason for the legend's initial strength was that to observers in the 1890s, it looked like the weevil *was* going to completely destroy the agricultural South. Cotton fields stretched across the region. The fleece was the foundation of the South's economic, social, and cultural organization. Few could escape cotton's grip. As soon as people began to understand how effective the boll weevil was against the plant, many understandably felt that the crop's demise was imminent.

Despite this fear that the weevil would end cotton production in the South, creating an agricultural, economic, and social revolution at the same time, it did not. For different reasons in different places, the rural order was not turned over by the boll weevil. By 1930, the pest had made its way across the entire region and the South was actually growing more cotton than when the weevil set out; landowners still enjoyed great advantages over the majority–African American labor force; and the region's credit system still forced thousands into debt and wedded the region to cotton production. This evidence of continuity over the period does not suggest, however, that the threat was not real, or that the battle to protect the status quo was not fought vigorously by different groups in different ways. The fight against the boll weevil was at its heart not a struggle between man and insect, but a contest between the region's landowners, tenant farmers, and rural educators, and it was never a foregone conclusion which side would win.

There were rare cases across the South when the boll weevil offered a course of change, but most areas eventually returned to pre-weevil conditions. Because of the insect, the sharecroppers on Johanna Reiser's Delta plantation enjoyed a brief stint renting their own land, but it was short-lived and exceptional. The diversification success in Coffee County, Alabama, likewise came as the result of the boll weevil, but it too failed. Of all the specific changes brought by the pest to the locales under study here, only the example of DPLC had any lasting legacy. The company, which owed its creation and development to the pest, still markets fast-growing cottonseed and produces tons of the white blossoms on its Delta plantations today. It serves not entirely as an example of continuity, however, because through the 1930s it remained a white-controlled plantation worked by a nearly all–African American labor force.

Throughout the rest of the South, the pest did not destroy all cotton equally. On soil that would allow a short growing season, farmers could modify their practices only modestly and still have a successful crop. On marginal lands, however, whether eroded, worn-out, hilly, soggy, rocky, or having some other undesirable attribute, the boll weevil often made it impossible to grow cotton, though land like this had rarely made farmers wealthy. In other places it was simply the timing of the weather that determined the success of the crop in the presence of the pest. As Arthur Raper's Greene County example shows, three consecutive years of rainy weather increased boll weevil populations and could produce unmatched destruction.

Not all of the reasons for the boll weevil's failures to bring a revolution to the South were the work of the weather or other natural phenomena, however. The extension service played a role in maintaining the economic and social arrangements of the South. Though the weevil's arrival to Texas and Louisiana helped to create the service itself, and it soon became a force on the rural scene, its agents never threatened to change the basic economic and social relationships in the countryside. Though its leaders believed that

diversification could be a solution to the boll weevil problem, it never suf-
ficiently supported the fundamental economic changes that a move from cotton necessitated.

As seen in the examples of the Mississippi Delta, southeastern Alabama,
and Georgia, the South's mono-crop system was at its base an economic
problem, not an agricultural one. Therefore, the boll weevil as strictly a farming issue mattered little; educators telling farmers how to more effectively till
their land, for example, was not going to improve the lot of southern farming
because it did not deal with the economic problems that forced farmers into
cotton (and debt) season after season. In southeast Alabama, for instance,
farmers' success growing peanuts became less important as the price of cotton
climbed. Even the threat of the boll weevil destroying an entire cotton crop
was pushed aside by the promise of high cotton prices. In Enterprise and
the rest of the South, diversification proponents tried to convince mid-sized
and small farmers to move away from cotton, but without creating structural
changes in the rural economy—credit, suppliers, and markets—there was no
real hope of delivering crop diversification to the agricultural South.

Some scholars have pointed to the rise of the extension service as one permanent change in the rural South produced by the boll weevil. While it is true
that federal agents moved into southern counties and became part of a rural
community's educational, political, and institutional lives in response to the
weevil threat, agents had little effect on the overall structures of these places.
This would change in the 1930s, when the federal government more drastically
and effectively dealt with the economic issues of farming and introduced real
change to the rural South. Rather than addressing the basic farming issues, as
agents had during the initial period of the boll weevil crisis, in the 1930s the
federal government began to address the fundamental problems in rural life.
The Agricultural Adjustment Act and a host of related New Deal policies put
the government at the heart of farmers' decisions about crop production and
labor, as well as rural bankers' and merchants' credit and marketing business.
These changes could have happened with the boll weevil—indeed they are
exactly what some reformers called for—but they did not.

So if the contention that the boll weevil was a destructive, revolutionary
force on the southern landscape is a myth, what was the pest's real legacy, and
what explains cotton's eventual demise throughout most of the South? In the
end, analyzing the effect of the boll weevil is as much about understanding
what change it didn't create as that which it did. While the beetle certainly
started a fight for the South's future, it didn't bring with it the revolution that
so many hoped or feared that it would.

TABLE 7. Cotton acreage by state, 1899–1929.

State	1899	1909	1919	1929
Alabama	3,202,135	3,730,892	2,628,160	3,566,494
Georgia	3,343,081	4,883,314	4,543,864	3,405,623
Louisiana	1,376,254	956,411	1,309,378	1,946,354
Mississippi	2,897,560	3,395,120	2,894,494	3,965,234
Texas	6,884,148	9,225,883	10,581,321	13,557,053

Data from Neil Fligstein, *Going North: Migration of Blacks and Whites from the South, 1900–1950* (New York: Academic Press, 1981), 89.

As Table 7 demonstrates, despite a brief downturn in cotton production in some states during the first few years of the boll weevil's presence, by 1930 cotton had rebounded, even surpassing pre-weevil levels in most places. A snapshot of that year would reveal a South much more similar to its state in 1900 than, say, 1945.

In 1930, land that had easily grown cotton before the weevil's arrival still produced the crop, and on less productive land farmers still struggled. In each of these places, however, the basic structure of the cotton society remained the same. The largest landowners had the easiest access to credit, relied heavily on a tenant workforce, and grew cotton with the lowest expenses. Mid-sized and small farmers still depended on some outside labor, struggled to afford the latest machines and services, and saw their bottom lines turn red as often as they showed a profit. For thousands of southern tenants, things were much different in 1930 than 1892, but only part of this was the result of the boll weevil. Though migration had been a constant in the life of landless laborers prior to the boll weevil, the pest had kept tenants in motion, and during the 1920s it helped some farmers to decide to move off the farm altogether, toward southern or northern cities. For the majority of tenant farmers, however, the arrival of the boll weevil had simply meant more internal migration from farm to farm. Those that gained access to land in the face of the pest recognized how rare their situation was and tried desperately to hold on to it.

The rural South's endemic problems did not arrive with the boll weevil nor did they end as farmers began to figure out ways to stop the pest. It was cotton, not its natural enemy, that made the South what it was. Even as late as 1936, the Southern Regional Committee of the Social Science Research Council could still claim that "the South is a land of cotton, and largely because of cotton the South is a region of problems." The group's report went on to describe how interrelated economic stagnation and enduring poverty were to the region's commitment to the staple:

> Cotton and high percentage of farm tenancy; cotton and a high ratio of Negro to white population; cotton and low family income; cotton and changing world-market conditions; cotton as king in the far-flung area of the South, and the appearance of new textiles; cotton and the one-crop system; cotton and soil wastage; cotton and a debtor economy—all of these combinations, and more, are the problems of the cotton economy.[5]

The boll weevil could have changed this situation, but it did not. In the end, one of the boll weevil's most important lost revolutions was that when cotton finally did die out on the South's marginal farm lands, the ground reverted to weeds and erosion gullies. Due to generations of farmers knowingly misusing the region's soil, much of the farmland was unable to support any kind of agriculture after cotton's demise. Rather than being turned over to other crops that might have assured the existence of a small farming class, the South's farmland became either corporate-owned and highly specialized, or basically wasteland.

It was not the boll weevil, but post–World War II changes in southern agriculture, which arrived as the result of the New Deal's structural attack on the rural South, technological innovation, and new national labor realities that finally brought revolution to southern farm life. Though many people in the early twentieth century thought it would be the boll weevil that brought these changes, and though many scholars have made that argument since, the pest had not revolutionized the rural South when it reached the Virginia coast.

The addendum to this modernization story often untold is that these quick changes hastened the destruction of the diversification dream itself. There would be few examples of balanced agriculture in the mid-twentieth century South. Farming became, and remains today, a specialized pursuit. A farming map of the South today shows islands of fruit, chickens, tobacco, soybeans, hogs, peanuts, and cotton. Though the dependence is no longer on a single crop across the entire region, it is still an adapted form of mono-crop agriculture.[6] Its continued presence in southern fields, however, is not the sole location of its legacy. In cities like Auburn, Athens, and Starkville, land grant colleges with multimillion dollar budgets owe a major part of their rise and modern power to the fight against the boll weevil. Likewise, many corporate farms and agribusinesses today owe their fortunes and land holdings to the insect.

If the basic agricultural and economic structures of the region survived the weevil's onslaught, then the pest's greatest legacy must be understood as a cultural one. The myth of the weevil's destruction of the plantation South has gained such a following that it is central to the way that southerners tell their history and view their region today. It has become as powerful a force in the region's own story as tenancy, farmland use, the Great Migration, and rural education. If not for this one little tiny insect, the story goes, the plantation South would have lasted forever. In the end, this explanation, though untrue, has become more powerful than the pest's reality.

Today, farmers, newspaper editors, and agricultural scientists love to celebrate not only the extinction of the boll weevil from southern cotton, as the NPR story did, but also the myth that the pest had forever changed the rural South. For example, a 2010 *New York Times* article about so-called "super weeds" spreading across the country quoted a Georgia cotton farmer who made the obvious comparison: "If we don't whip this thing, it's going to be like the boll weevil did to cotton . . . It will take it away."[7]

It is not simply the farmers who embrace that myth. In various small southern towns the boll weevil is emblazoned on city monuments and football helmets. Southerners can eat at weevil-themed restaurants and buy knickknacks featuring images of the bug with the phrase "Lest we forget." The boll weevil pervades southern culture; it is a crucial component of the larger personal, cultural, and economic history that southerners tell of their region. But it is a force whose power came from a mix of natural and human factors, forged during its first thirty years in the United States.

NOTES

INTRODUCTION

1. Theodore Roosevelt, "Annual Message to Congress," December 6, 1904, reprinted in *New York Times*, December 8, 1904; Georgia State Board of Entomology, "Regulations of the Georgia State Board of Entomology, Relative to the Quarantine Against the Mexican Cotton Boll Weevil," December 24, 1904, in Duggar Family Papers, box 7, folder 80, Auburn University Special Collections.

2. O. F. Cook, "An Enemy of the Cotton Boll Weevil" *Science*, New Series, vol. 19, no. 492 (June 3, 1904): 862–64.

3. "Scientific Notes and News," *Science*, New Series, 18, no. 463 (November 13, 1903), 640 (italics added).

4. Seaman Knapp quoted in Joseph Cannon Bailey, *Seaman A. Knapp: Schoolmaster of American Agriculture* (New York: Columbia University Press, 1945), 169.

5. The maps actually showed only the farthest extent of the boll weevil's movement at the close of a season. The weevil population might have been inconsequentially small at its farthest migration point, and there were vast areas within the "weevil territory" that suffered little from the pest, but the maps gave the impression that the insect had completely covered cotton land to the outer markers.

6. Since the weevil destroys cotton squares on the plant, not bales of picked cotton, figures that suggest crop loss in bales or by weight are a bit problematic. Dickerson et al. gathered acreage crop loss figures, converted that to bales based on a contemporary formula of bales per acre, assumed all bales were five hundred pounds, and multiplied the total pound loss by the average contemporary price of cotton in cents per pound. These figures are not, therefore, perfect, but are the best data extant for annual crop loss and value, especially for this early period. Willard A. Dickerson et al., eds., *Boll Weevil Eradication in the United States through 1999*, Cotton Foundation Reference Book Series, no. 6 (Memphis: Cotton Foundation Publisher, 2001), 590.

7. Harris Dickson, *The Story of King Cotton* (New York: Funk & Wagnalls Company, 1937), 96.

8. William Atherton Du Puy, "The Insects Are Winning," *Harper's Magazine* (March 1925): 436.

9. Leland O. Howard, *The Insect Menace* (New York: Century Company, 1931), 317.

10. I define "myth" as a misrepresentation told and retold, wittingly and unwittingly, by people who can gain from it.

11. Ted Steinberg offers a compelling analysis of man's place in defining natural disasters in *Acts of God: The Unnatural History of Natural Disaster in America* (New York: Oxford University Press, 2000).

12. United States Department of Agriculture, "The Man Who Works With His Hands: Address of President Roosevelt at the Semi-Centennial Celebration of the Founding of Agricultural Colleges in the United States, at Lansing, Michigan, May 31, 1907," circular no. 24 (July 1, 1907), 10.

13. There have been two dissertations devoted to the boll weevil. See Douglas Helms, "Just Lookin' for a Home: The Cotton Boll Weevil and the South" (Ph.D. dissertation, Florida State University, 1977), and James C. Giesen, "The South's Greatest Enemy? The Cotton Boll Weevil and Its Lost Revolution, 1892–1930" (Ph.D. dissertation, University of Georgia, 2004).

14. For examples of treatment of the boll weevil that focuses on economic disruption, the growth of the extension service, and changing relationships between farmers and the state, see Helms, "Just Lookin' for a Home"; Fabian Lange, Alan Olmstead, and Paul Rhode, "The Impact of the Boll Weevil, 1892–1932," *Journal of Economic History* 69, no. 3 (September 2009): 685–718; Pete Daniel, *Breaking the Land: The Transformation of Cotton, Tobacco, and Rice Cultures since 1880* (Urbana: University of Illinois Press, 1985); Robert Higgs, "The Boll Weevil, The Cotton Economy, and Black Migration: 1910–1930," *Agricultural History* 50 (April 1976): 335–50; Kathryn Holland Braund, "'Hog Wild' and 'Nuts': Billy Boll Weevil Comes to the Alabama Wiregrass," *Agricultural History* 63 (1989): 15–39; Kent Osband, "The Boll Weevil Versus 'King Cotton,'" *Journal of Economic History* 45, no. 3 (1985): 627–43; Arvarh E. Strickland, "The Strange Affair of the Boll Weevil: The Pest as Liberator," *Agricultural History* 68 (1994): 157–68; Charles S. Aiken, *The Cotton Plantation South since the Civil War* (Baltimore: Johns Hopkins University Press, 1998).

15. Karen Brown, "Political Entomology: The Insectile Challenge to Agricultural Development in the Cape Colony, 1895–1910," *Journal of Southern African Studies* 29, no. 2. (June 2003): 529.

16. Braund, "'Hog Wild' and 'Nuts,'" discusses the background for building the boll weevil monument but doesn't investigate the power of the monument's message, either to farmers in southeast Alabama or the rest of the South. Joshua Blu Buhs discusses the persuasive power of images of the fire ant in his book on the insect, but this cultural component is not central to his argument. See Blu Buhs, *The Fire Ant Wars: Nature, Science, and Public Policy in Twentieth-Century America* (Chicago: University of Chicago Press, 2004).

17. Donald Worster, *Dust Bowl: The Southern Plains in the 1930s* (New York: Oxford University Press, 1979). See also William Cronon, "A Place for Stories: Nature, History, and Narrative," *Journal of American History* 78, no. 4 (March 1992): 1347–76.

18. The field of southern environmental history grew exponentially in the first decade of

the twenty-first century, though what the field has brought to our collective understanding of southern agriculture is still somewhat limited. There are important exceptions to this, however. Mart A. Stewart has made the most convincing argument about the intersections of these two fields, and has contributed some of the best work on southern farming from an environmental history perspective. See Stewart, "If John Muir Had Been an Agrarian: American Environmental History West and South," *Environment and History* 11, no. 2 (2005); and Stewart, *"What Nature Suffers to Groe": Life, Labor, and Landscape on the Georgia Coast, 1680 –1920* (Athens: University of Georgia Press, 1996). Other important southern environmental history includes Lynne A. Nelson, *Pharsalia: An Environmental Biography of a Southern Plantation, 1780 –1880* (Athens: University of Georgia Press, 2007); Jack Temple Kirby, *Mockingbird Song: Ecological Landscapes of the South* (Chapel Hill: University of North Carolina Press, 2006); Mikko Saikku, *This Delta, This Land: An Environmental History of the Yazoo-Mississippi Floodplain* (Athens: University of Georgia Press, 2005); Carville Earle, "The Myth of the Southern Soil Miner: Macrohistory, Agricultural Innovation, and Environmental Change," in Donald Worster, ed., *The Ends of the Earth: Perspectives on Modern Environmental History* (Cambridge: Cambridge University Press, 1988); Mark Hersey, "Hints and Suggestions to Farmers: George Washington Carver and Rural Conservation in the South," *Environmental History* 11 (April 2006): 239–68; Albert G. Way, "Burned to Be Wild: Herbert Stoddard and the Roots of Ecological Conservation in the Southern Longleaf Pine Forest," *Environmental History* 11 (July 2006): 500–526. The older work within southern history that focused at least in part on what we would today call the environment should remain relevant to newer perspectives, and at the very least deserves greater mention in the emerging environmental literature of the South. Some of this work is quite old, evidence in itself of southern scholars' long-running attention to the natural landscape of the region. These include far too many to list here, but those most germane to the argument of this book include Charles S. Aiken, *The Cotton Plantation South since the Civil War* (Baltimore: Johns Hopkins University Press, 1998); Daniel, *Breaking the Land*; Arthur Raper, *Preface to Peasantry: A Tale of Two Black Belt Counties* (Chapel Hill: University of North Carolina Press, 1936); Rupert Vance, *Human Factors in Cotton Culture: A Study in the Social Geography of the American South* (Chapel Hill: University of North Carolina Press, 1929); C. Vann Woodward, *Origins of the New South, 1877–1913* (Baton Rouge: Louisiana State University Press, 1951); Steven Hahn, "Hunting, Fishing, and Foraging: Common Rights and Class Relations in the Postbellum South," *Radical History Review* 26 (October 1982): 37–64.

CHAPTER 1

1. Seaman A. Knapp, "An Address at the Anti–Boll Weevil Conference for the Southeastern States" (Atlanta, 1910), 6, as quoted in Joseph Cannon Bailey, *Seaman A. Knapp: Schoolmaster of American Agriculture* (New York: Columbia University Press, 1945), 169–71. Harris Dickson retold an almost identical story: "One farmer sprang up at a scare meeting and shouted, 'Tain't no sense tryin' to fight that devil. I corked up a lot of 'em in pure alcohol and kept 'em for two hours. They come out staggering drunk, and with a mighty good appetite. Then I sealed 'em in a tin can an' throwed 'em in the fire. When the can melted, them red-hot

bugs flew out and burnt my barn." Dickson, *The Story of King Cotton* (New York: Funk & Wagnalls Company, 1937), 97.

2. Knapp, "Farmers' Cooperative Demonstration Work and Its Results" (Richmond, VA, 1906), in Bailey, *Seaman A. Knapp*, 169.

3. More description of the social and economic effects of cotton's arrival to southern Texas follows in chapter 2. For background on the extension of railroads and cotton into southern Texas and the consequent identification of the boll weevil there, see Neil Foley, *The White Scourge: Mexicans, Blacks, and Poor Whites in Texas Cotton Culture* (Berkeley: University of California Press, 1997); David Montejano, *Anglos and Mexicans in the Making of Texas, 1836–1986* (Austin: University of Texas Press, 1987); E. Dwight Sanderson, "The Mexican Cotton Boll Weevil," Texas Agricultural Experiment Station, Entomological Department, circular no. 1 (February 15, 1903); John Solomon Otto, *Southern Agriculture during the Civil War Era, 1860–1880* (Westport, CT: Greenwood Press, 1994), 16, 86; Rebecca Sharpless, *Fertile Ground, Narrow Choices: Women on Texas Cotton Farms, 1900–1940* (Chapel Hill: University of North Carolina Press, 1999), 4; John S. Spratt, *The Road To Spindletop: Economic Change in Texas, 1875–1901* (Austin: University of Texas Press, 1955), 62; Roger L. Ransom and Richard Sutch, *One Kind of Freedom: The Economic Consequences of Emancipation*, 2nd ed. (Cambridge: Cambridge University Press, 2001), 172; Earle B. Young, *Galveston and the Great West* (College Station: Texas A&M University Press, 1997), 62.

4. Antonio A. Guerra, "Seasonal Boll Weevil Movement between Northeastern Mexico and the Rio Grande Valley of Texas, USA," *Southwestern Entomologist* 14, no. 4 (December 1988): 261; W. H. Cross, M. J. Lukefahr, P. A. Fryxell, and H. R. Burke, "Host Plants of the Boll Weevil," *Environmental Entomology* 4, no. 1 (February 1975): 19; Kyung Seok Kim and Thomas W. Sappington, "Boll Weevil (Anthonomus grandis Boheman) (Coleoptera: Curculionidae) Dispersal in the Southern United States: Evidence from Mitochondrial DNA Variation," *Environmental Entomology* 33, no. 2 (2004): 457; Kyung Seok Kim and Thomas W. Sappington, "Genetic Structuring of Boll Weevil Populations in the U.S. Based on RAPD Markers," *Insect Molecular Biology* 13 (2004): 300; Frederick L. Lewton, "Cienfuegosia Drummondii, a Rare Texas Plant," *Bulletin of the Torrey Botanical Club* 37 (1910): 473–74; Donovan Stewart Correll and Marshall Conring Johnston, *Manual of the Vascular Plants of Texas* (Renner: Texas Research Foundation, 1970), 1033.

5. Guerra, "Seasonal Boll Weevil Movement Between Northeastern Mexico and the Rio Grande Valley of Texas, USA," *Southwestern Entomologist* 14, no. 4 (December 1988): 261; Cross, et al., "Host Plants of the Boll Weevil," *Environmental Entomology* 4, no. 1 (February 1975): 19; Kim and Sappington, "Boll Weevil (Anthonomus grandis Boheman) (Coleoptera: Curculionidae) Dispersal in the Southern United States: Evidence from Mitochondrial DNA Variation," *Environmental Entomology* 33, no. 2 (2004): 457.

6. C. H. DeRyee to the Commissioner of Agriculture, October 3, 1894, as quoted in John Douglas Helms, "Just Looking for a Home: The Cotton Boll Weevil and the South" (Ph.D. dissertation, Florida State University, 1977), 3–4.

7. Horace R. Burke, Wayne E. Clark, James R. Cate, and Paul A. Fryxell, "Origin and Dispersal of the Boll Weevil," *Bulletin of the Entomological Society of America* 32, no. 4 (Winter 1986): 229; W. H. Johnson, *Cotton and Its Production* (London: Macmillan and Co., 1926),

473; Evans, "Texas Agriculture, 1880–1930," 60; Helms, "Just Looking for a Home," 4; Rogers McVaugh, *Edward Palmer: Plant Explorer of the American West* (Norman: University of Oklahoma Press, 1956), 81; Walter D. Hunter and Warren E. Hinds, "The Mexican Cotton-Boll Weevil: A Summary of the Investigation of this Insect up to December 31, 1911," Senate Document 305, 62nd Congress, Second Session, 15.

8. Acting Secretary to C. H. DeRyee, October 26, 1894, as quoted in Helms, "Just Looking for a Home," 5; Helms, "Just Looking for a Home," 9–12. For background on the importance of Leland Howard and his career fighting all kinds of insects, see James E. McWilliams, *American Pests: The Losing War on Insects from Colonial Times to DDT* (New York: Columbia University Press, 2008), chapter 5. See also Hae-Gyung Geong, "Exerting Control: Biology and Bureaucracy in the Development of American Entomology, 1870–1930" (Ph.D. dissertation, University of Wisconsin–Madison, 1999).

9. C. H. Tyler Townsend, "Report on the Mexican Cotton-Boll Weevil in Texas," *Insect Life* 7, Division of Entomology, U.S. Department of Agriculture (Washington, D.C.: March 1895), 295–301; Helms, "Just Looking for a Home," 9–12.

10. The description of the boll weevil offered in this paragraph and in those that immediately follow relies on recent entomological research to paint a more accurate description of the reality of the pest than was fully known in the 1890s. Mississippi State University Cooperative Extension Service, "Management and Control of Boll Weevils," publication 1830 (1992), 3.

11. Thomas F. Leigh, Steven H. Roach, and Theo F. Watson, "Biology and Ecology of Important Insect and Mite Pests of Cotton," in *Cotton Insects and Mites: Characterization and Management*, edited by Edgar G. King, Jacob R. Phillips, and Randy J. Coleman (Memphis: Cotton Foundation Publisher, 1996), 18; T. B. Davich, "Introduction," in P. P. Sikorowski, J. G. Griffin, J. Roberson, and O. H. Lindig, *Boll Weevil Mass Rearing Technology* (Jackson: University Press of Mississippi, 1984), 3; Donald J. Borror and Dwight M. DeLong, *An Introduction to the Study of Insects*, 3rd ed. (New York: Holt, Rinehart and Winston, 1971), 357; R. A. Crowson, *The Biology of the Coleoptera*, U.S. ed. (London: Academic Press, 1981), 613; R. F. Chapman, *The Insects: Structure and Function*, 4th ed. (Cambridge: Cambridge University Press, 1998), 70–71, 74, 708.

12. Mississippi State University Cooperative Extension Service, "Management and Control of Boll Weevils," 2; Robert E. Pfadt, "Insect Pests of Cotton," in Pfadt, ed., *Fundamentals of Applied Entomology*, 3rd ed. (New York: Macmillan Publishing Co, Inc., 1978), 383, 386; Leigh, Roach, and Watson, "Biology and Ecology of Important Insect and Mite Pests of Cotton," 18–19; S. M. Greenberg, T. W. Sappington, D. W. Spurgeon, and M. Sétamou, "Boll Weevil (Coleoptera: Curculionidae) Feeding and Reproduction as Functions of Cotton Square Availability," *Environmental Entomology* 32, no. 3 (June 2003): 699.

13. Mississippi State University Cooperative Extension Service, "Management and Control of Boll Weevils," 2–4; Pfadt, "Insect Pests in Cotton," 383, 386; John M. Munro, *Cotton*, 2nd ed. (New York: Longman Scientific and Technical, 1987), 151.

14. Mississippi State University Cooperative Extension Service, "Management and Control of Boll Weevils," 2–4; J. R. Bradley, Jr., "Major Developments in Management of Insect and Mite Pests in Cotton," in *Cotton Insects and Mites: Characterization and Management*,

eds. Edgar G. King, Jacob R. Phillips, and Randy J. Coleman, 1–2; May R. Barenbaum, *Ninety-nine More Maggots, Mites, and Munchers* (Urbana: University of Illinois Press, 1993), 34–37; Helms, "Just Looking for a Home," 7; Pfadt, "Insect Pests in Cotton," 381, 383; W. B. Mercier and H. E. Savely, *The Knapp Method of Growing Cotton* (New York: Doubleday, Page & Company, 1913), 108.

15. Thomas F. Leigh, Steven H. Roach, and Theo F. Watson, "Biology and Ecology of Important Insect and Mite Pests of Cotton," in *Cotton Insects and Mites: Characterization and Management*, eds. Edgar G. King, Jacob R. Phillips, and Randy J. Coleman (Memphis: Cotton Foundation Publisher, 1996), 19; Mississippi State University Cooperative Extension Service, "Management and Control of Boll Weevils," 2; Pfadt, ed., "Insect Pests in Cotton," 383, 386.

16. C. H. Tyler Townsend, "Report on the Mexican Cotton-Boll Weevil in Texas," *Insect Life* 7, Division of Entomology, U.S. Department of Agriculture (Washington, D.C.: March 1895), 295–301.

17. Townsend, "Report on the Mexican Cotton-Boll Weevil in Texas," 305; Helms, "Just Looking for a Home," 9–12; E. Dwight Sanderson, "The Mexican Cotton Boll Weevil," Texas Agricultural Experiment Stations, Entomological Department, circular no. 1 (February 15, 1903); Leland O. Howard, *The Insect Menace* (New York: Century Company, 1931) 315. For background on the debates between cultural methods and chemical control, see Geong, "Exerting Control: Biology and Bureaucracy in the Development of American Entomology, 1870–1930," and McWilliams, *American Pests*.

18. Townsend, "Report on the Mexican Cotton-Boll Weevil in Texas," 304–6; Helms, "Just Looking for a Home," 9–12; E. Dwight Sanderson, "The Mexican Cotton Boll Weevil," Texas Agricultural Experiment Stations, Entomological Department, circular no. 1 (February 15, 1903); Leland O. Howard, *The Insect Menace* (New York: Century Company, 1931), 315.

19. Farmers in the southern tip of Texas still relied more on fruit growing and livestock grazing than cotton at this point. Townsend, "Report on the Mexican Cotton-Boll Weevil in Texas," 307–8.

20. There would be several suggestions made to enact similar cotton-free zones during the first decade of the pest's spread across the South. Hunter and the USDA would eventually decide against supporting any of these proposals because they were each deemed unfeasible for both entomological and nonentomological reasons. Not only were they expensive, the bans would have been impossible to enforce. The pests could travel on their own (it would be hard to police borders for individual weevils) or be shipped unintentionally along with dozens of agricultural products. Getting different communities and states to agree on the place and timing of the belts would also prove to be an insurmountable obstacle. Helms, "Just Looking for a Home," 56–57.

21. A. S. Packard, Jr., *The Injurious Insects of the West: A Report on the Rocky Mountain Locust and Other Insects Now Injuring or Likely to Injure Field and Garden Crops in the Western States and Territories* (Salem: S. E. Cassino, 1877); Jeffrey A. Lockwood, *Locust: The Devastating Rise and Mysterious Disappearance of the Insect That Shaped the American Frontier* (New York: Basic Books, 2004). For more background on the federal government's battles against insect pests prior to the boll weevil see McWilliams, *American Pests*.

22. Elizabeth Sanders, *Roots of Reform: Farmers, Workers, and the American State, 1877–*

1917 (Chicago: University of Chicago Press, 1999), 314–18. For additional background on the rise of state experiment stations, see Alan I. Marcus, *Agricultural Science and the Quest for Legitimacy* (Ames: Iowa State University Press, 1985); Roy V. Scott, *The Reluctant Farmer: The Rise of Agricultural Extension to 1914* (Urbana: University of Illinois Press, 1970); Alfred Charles True, *A History of Agricultural Education in the United States, 1785–1925* (New York: Arno Press, 1969).

23. As mentioned, a longer discussion of the cotton labor system and the connection between cotton farmers and international market forces follows in chapter two. There is a vast literature on the development of the sharecropping system after the Civil War, the evolution of the global cotton economy, and its effect on southern farmers, though very little of it takes into account disruptive forces like the boll weevil. See Harold Woodman, *King Cotton and His Retainers: Financing & Marketing the Cotton Crop of the South, 1800–1925* (Lexington: University of Kentucky Press, 1968); Woodman, *New South, New Law: The Legal Foundations of Credit and Labor Relations in the Postbellum Agricultural South* (Baton Rouge: Louisiana State University Press, 1995); Ransom and Sutch, *One Kind of Freedom*; Gavin Wright, *Old South, New South: Revolutions in the Southern Economy since the Civil War* (New York: Basic Books, 1986).

24. John D. Hicks, *The Populist Revolt* (Minneapolis: University of Minnesota Press, 1931), 47, as quoted in Spratt, *Road To Spindletop*, 69.

25. Bailey, *Seaman A. Knapp*, 207. For an example of an early, confusing bulletin sent to farmers, see Texas Agricultural Experiment Station, "Sundry Brief Articles," bulletin no. 37 (December 1895).

26. *Eighth Annual Report of the Texas Agricultural Experiment Stations for 1895* (College Station: Agricultural and Mechanical College of Texas, 1896); Helms, "Just Looking for a Home," 9–14, 18.

27. Eighth Annual Report of the Texas Agricultural Experiment Stations for 1895, 756–57.

28. Congress, Senate, Senator Blanchard of Louisiana, 54th Cong., 1st sess., *Congressional Record* (February 17, 1896): 1782–83.

29. Ibid (Italics added).

30. Ibid.; Sanders, *Roots of Reform*.

31. Frank Wagner, "The Boll Weevil Comes to Texas," *Occasional Papers* 6 (Corpus Christi: Friends of the Corpus Christi Museum, June 1980), 10.

32. See introduction for an explanation of why these estimates of expected "bales destroyed" are flawed.

33. D. F. Houston, "Cotton and the General Agricultural Outlook," *Publications of the American Economic Association*, 3rd series, vol. 5, no. 1, Papers and Proceedings of the Sixteenth Annual Meeting, Part I, New Orleans, LA., December 29–31, 1903 (February 1904), 114–16.

34. Bailey, *Seaman A. Knapp*, title page; C. Vann Woodward, *Origins of the New South, 1877–1913* (Baton Rouge: Louisiana State University Press, 1954), 119.

35. Italics added. Jackson Davis, "An Experiment in Agricultural Education," in "Silver Anniversary Cooperative Demonstration Work, 1903–1928," Proceedings of the Anniversary

Meeting Held at Houston, Texas, February 5–7, 1929 (College Station: Agricultural and Mechanical College of Texas), 38, hereafter "Silver Anniversary Proceedings." Charles Brough, "The American Farmer and Agriculture of the Future," in "Silver Anniversary Proceedings," 54; Walter Hines Page, as quoted by Jackson Davis, "Silver Anniversary Proceedings," 38; R. R. Moton, "Extension Work and the Negro," in "Silver Anniversary Proceedings," 51; Davis, "An Experiment in Agricultural Education," in "Silver Anniversary Proceedings," 38.

36. Cline, "Seaman Asahel Knapp, 1833–1911," 333–34; Oscar B. Martin, *The Demonstration Work: Dr. Seaman A. Knapp's Contribution to Civilization* (Boston: Stratford Company, 1926), 2; Bailey, *Seaman A. Knapp*, 98.

37. Cline, "Seaman Asahel Knapp, 1833–1911," 334–35; Williamson, *Origin and Growth*, 42, 44.

38. In fact, Louisiana had become the nation's leading rice producer after Knapp's first five years in the state. Cline, "Seaman Asahel Knapp," 336–38; Williamson, *Origin and Growth*, 49; Martin, *The Demonstration Work*, 8–10; Woodward, *Origins of the New South*, 119.

39. Williamson, *Origin and Growth*, 45; Woodward, *Origins of the New South*, 409.

40. Martin, *The Demonstration Work*, 1.

41. Bailey, *Seaman A. Knapp*, 145–47; Paul D. Peterson Jr. and C. Lee Campbell, "Beverly T. Galloway: Visionary Administrator," *Annual Review of Phytopathology* 35 (September 1997), 32.

42. Rasmussen, *Readings in the History of American Agriculture*, 178–79; Bailey, *Seaman A. Knapp*, 149–52l "Green, Edward Howland Robinson," The Handbook of Texas Online, http://www.tshaonline.org/handbook/online/articles/fgr33, accessed October 26, 2010; "Texas Midland Railroad," The Handbook of Texas Online, http://www.tshaonline.org/handbook/online/articles/eqt22, accessed October 26, 2010.

43. Rasmussen, *Readings in the History of American Agriculture*, 178–79; Bailey, *Seaman A. Knapp*, 149–52.

44. Jack Stoltz, "The Porter Demonstration Farm," *East Texas Historical Journal* 1992 30(1): 16; Davis, in "Silver Anniversary Proceedings," 39, 41.

45. Stoltz, "The Porter Demonstration Farm," 17.

46. The Porter Farm was technically owned by Walter Porter's elderly father, J. B. Porter. Stoltz, "The Porter Demonstration Farm," 17; Williamson, *Origin and Growth*, 49–51; Bailey, *Seaman A. Knapp*, 152–55.

47. The first wave of histories written about Knapp and the Porter Farm latched on to this idea that the boll weevil had been defeated there. Knapp biographer Joseph Cannon Bailey wrote, "The project on the Porter farm at Terrell was completed in a crisis of such magnitude that, in Texas, seemed to threaten the extinction of all cotton culture." Bailey, *Seaman A. Knapp*, 152–55, quote on 161; Williamson, *Origin and Growth*, 49–51; Stoltz, "The Porter Demonstration Farm," 18.

48. Williamson, *Origin and Growth*, 49–51.

49. Hearings before the Committee on Agriculture on Bills Having for Their Object the Eradication of the Cotton-Boll Weevil and Other Insects and Diseases Injurious to Cotton, Congress, House, 58th Cong., 2nd sess., 1904: 9, 35; Howard, *The Insect Menace*, 68–69.

50. Hearings before the Committee on Agriculture on Bills Having for Their Object the

Eradication of the Cotton-Boll Weevil and Other Insects and Diseases Injurious to Cotton, Congress, House, 58th Cong., 2nd sess., 1904: 36; Congress, House, Representative Hemenway of Indiana, 58th Cong., 2nd sess., *Congressional Record* (January 7, 1904): 543–44.

51. Congress, House, Representative Burleson of Texas, 58th Cong., 2nd sess., *Congressional Record* (January 8, 1904): 569–72.

52. Gillett's comments spoke to the heart of the local vs. national debate: "I suppose that what will govern Congress in appropriating against any such a pest is whether it is a local or a national calamity." "I suggest," he quickly added "that it is going to be very hard ever to draw that line." Congress, House, Representative Gillett of Massachusetts, 58th Cong., 2nd sess., *Congressional Record* (January 8, 1904): 569, 570.

53. Congress, House, Representative Robinson of Arkansas, 58th Cong., 2nd sess., *Congressional Record* (January 8, 1904): 572; Congress, House, Representative Slayden of Texas, 58th Cong., 2nd sess., *Congressional Record* (January 8, 1904): 572–73.

54. Martin, *The Demonstration Work*, 4; Scott, *Reluctant Farmer*, 213.

55. Helms, "Just Looking for a Home," 168.

CHAPTER 2

1. Davis, "An Experiment in Agricultural Education," in "Silver Anniversary Collection," 40. The photograph later appeared in Martin, *The Demonstration Work*, 33.

2. Davis, "An Experiment in Agricultural Education," in "Silver Anniversary Collection," 40.

3. Debra A. Reid, "Reaping a Greater Harvest: African Americans, Agrarian Reform, and the Texas Agricultural Extension Service" (Ph.D. dissertation, Texas A&M University, 2000), 50–52; Reid, *Reaping a Greater Harvest: African Americans, the Extension Service, and Rural Reform in Jim Crow Texas* (College Station: Texas A&M Press, 2007), 6–8; Reid, "Racism and Sexism in Rural Texas: The Contested Nature of Progressive Rural Reform, 1870s-1910s," in Reid, ed., *Seeking Inalienable Rights: Texans and Their Quests for Justice* (College Station: Texas A&M Press, 2009), 37–57, esp. 55n19.

4. Not all historians have written glowingly of Knapp, and certainly not of the cooperative extension service that grew out of the Porter Farm. Beginning in the 1970s but reaching full steam in the late 1980s scholars grew more critical of Knapp and indeed all government farm experts and their effects on farm life. The most important of this work has come from Pete Daniel, who has concentrated on the government's missteps with farmers in the majority of his work. See Daniel, *Breaking the Land*; Daniel, *Lost Revolutions*; Daniel, *Toxic Drift*. Indeed, nearly all works on southern farm life since 1980 have contained some criticism of government programs. See, for instance, Fite, *Cotton Fields No More*; Kirby, *Rural Worlds Lost*; Ayers, *Promise of the New South*; Reid, *Reaping a Greater Harvest*.

5. On historians' dependence on a few stock stories of sharecropping see James C. Giesen, "Creating 'Nate Shaw': The Making and Re-making of *All God's Dangers*," in Richard Godden and Martin Crawford, eds., *Reading Southern Poverty Between the Wars* (Athens: University of Georgia Press, 2006).

6. My perspective on tenant movement and decisions is influenced by Peter Coclanis and

Bryant Simon, "Exit, Voice, and Loyalty: African American Strategies for Day-to-Day Existence/Resistance in the Early-Twentieth-Century Rural South," in R. Douglas Hurt, ed. *African American Life in the Rural South, 1900–1950* (Columbia: University of Missouri Press, 2003),189–209.

7. Historians have written a great deal about the origins and mechanics of the southern tenant system and sharecropping in particular. See Stephen J. DeCanio, *Agriculture in the Postbellum South: The Economics of Production and Supply* (Cambridge: MIT Press, 1974); Edward Ayers, *Promise of the New South* (New York: Oxford University Press, 1992); Pete Daniel, *The Shadow of Slavery: Peonage in the South, 1901–1969* (Urbana: University of Illinois Press, 1972); William Cohen, *At Freedom's Edge: Black Mobility and the Southern White Quest for Racial Control, 1861–1915* (Baton Rouge: Louisiana State University Press, 1991); Roger L. Ransom, and Richard Sutch, *One Kind of Freedom: The Economic Consequences of Emancipation*, 2nd ed. (Cambridge: Cambridge University Press, 2001), George Brown Tindall, *The Emergence of the New South, 1913–1945* (Baton Rouge: Louisiana State University Press, 1967).

8. Sharpless, *Fertile Ground, Narrow Choices*, 7–8; Neil Foley, *The White Scourge: Mexicans, Blacks, and Poor Whites in Texas Cotton Culture* (Berkeley: University of California Press, 1997), 42; Alwyn Barr, *Black Texans: A History of African Americans in Texas, 1528–1995*, 2nd ed. (Norman: University of Oklahoma Press, 1996), 88–89.

9. The Division of Statistics figured these averages monthly, despite the fact that the majority of southern farm labor worked under non-fixed wage systems; workers did not get paid each month. In 1890, as cotton fever was still spreading through eastern and southern Texas, the labor supply dried up. That year Texas agricultural workers were earning $13.30 per month, with board, according to the USDA's Division of Statistics; this was almost one dollar more than the national average. By the end of the decade, however, Texas's farm workers earned less money ($12.94) and had slipped a dollar below the national average. On the surface, this loss of wages suggests a rise in the labor supply, but the reality was a bit more complex. Compared to the wage averages of other states in the cotton belt, Texas landowners still paid the best wages in the South. In Georgia, for instance, farm workers earned only $8.05 per month in 1899, slightly more than *half* the national wage average of $14.07. While California farm labor was earning $25.64 per month with board that year, the average of the southern cotton belt states was a paltry $9.45. As for the division's breakdown of race, it is unclear whether Mexican workers were considered "colored" in Texas, though they were considered so in New Mexico and Arizona. John Hyde, *Wages of Farm Labor in the United States: Results of Eleven Statistical Investigations, 1866–1899* (Washington, D.C.: United States Department of Agriculture, Division of Statistics, 1901), 11.

10. Paul S. Taylor, *An American-Mexican Frontier: Nueces County, Texas* (Chapel Hill: University of North Carolina Press, 1934), 278; Foley, *The White Scourge*, 28–29.

11. Foley, *The White Scourge*, 9; Taylor, *An American-Mexican Frontier*, 300; Montejano, *Anglos and Mexicans in the Making of Texas*, 73. One irony of southern farm life was that cotton pickers were divided physically, socially, and economically from those that did not have to pick cotton, yet these groups were so closely connected to the livelihood of the region that landowners were obsessed with these connections themselves. There was a stigma attached to anyone

who picked cotton in the South, albeit for the vast majority of rural dwellers, white, black or Mexican, everyone picked cotton when it was time. Adding the boll weevil to the equation in the 1890s further complicated the already uneven race and class landscape. For evidence of the stigma attached to whites picking cotton, see Evans, "Texas Agriculture," 42.

12. U. C. Loftin, "Living with the Boll Weevil for Fifty Years," Annual Report of the Board of Regents of the Smithsonian Institution for 1945 (Washington, D.C., 1945), 273, as quoted in Ian R. Manners, "The Persistent Problem of the Boll Weevil: Pest Control in Principle and in Practice," *Geographical Review*, vol. 69, no. 1 (January 1979): 25.

13. Gloster (Mississippi) *Herald*, October 23, 1903, 4; W. D. Hunter, "The Status of the Mexican Cotton Boll-Weevil in the United States in 1903," in *Yearbook of the United States Department of Agriculture 1903* (Washington, D.C.: Government Printing Office, 1904), 207.

14. Douglas Helms, "Technological Methods for Boll Weevil Control," in George L. Robson Jr. and Roy V. Scott, eds., *Southern Agriculture since the Civil War: A Symposium* (Santa Barbara: McNally & Lotin, West, 1979), 290; Katie Dickie Stavinoha and Lorie A. Woodward, "Texas Boll Weevil History," in Dickerson et al., *Boll Weevil Eradication in the United States*, 461.

15. L. E. Perrin, "Cooperative Demonstration Days," in "Silver Anniversary Proceedings," 81. The phrase "pick squares" refers to the method sometimes advanced in the early years by state experts of walking through the cotton row by row and picking up any squares that had been punctured by a boll weevil and fallen to the ground. Weevil eggs hatched from the squares and farmers believed that they could reduce the number of the insect by collecting and burning these squares before they hatched. The laborious work was rarely worth the effort, since a few missed squares could mean the birth of thousands of weevils by the end of the season.

16. Ibid.

17. Perrin, "Cooperative Demonstration Days," in "Silver Anniversary Proceedings," 78. It is hard to verify this story. There is no B. B. Sochon in St. Landry Parish listed in either the 1900 or 1910 census. This neither precludes the possibility that he worked for Stelly in 1908 nor proves that Perrin's story of Sochon moving after the season is true. There were two different heads of household listed in the parish for Louis D. Stelly listed as farmers, though neither had any nonfamily farm labor living with him. U.S. Department of Commerce, *Twelfth Census of the United States: 1900* (Washington, D.C.: 1902); U.S. Department of Commerce, *Thirteenth Census of the United States: 1910* (Washington, D.C.: 1912).

18. Southern cities' populations increased steadily as the boll weevil spread across the region, in part because farmers left the land in search of better work and more fulfilling lives in urban areas.

19. One point often missed in the discussion of migration patterns of southern farmers is the extent to which landowners were tied to one place. Landowners were more tied to land than tenants were. Small farm owners relied on extra-familial labor to plant, chop, and pick cotton, and the economic structure of the cotton economy—the need for credit—chained these landowners to their own farms and to the merchants and banks that supplied them capital every season. As economic historian Gavin Wright has shown, even prosperous landowners had to offer property as collateral to secure relatively small loans so that the owner could buy seed and

tools each spring. These debts tied owners to the land as much or even more than personal debts tied workers to a particular place. During disruptions in farming, like the advent of the boll weevil, it could be an advantage to not own land, so that one could move away from the insect invader. Wright, *New South, Old South*, 112.

20. Robert Higgs, "The Boll Weevil, the Cotton Economy, and Black Migration, 1910–1930" *Agricultural History* 50 (1976): 335–50; Arvarh E. Strickland, "The Strange Affair of the Boll Weevil: The Pest as Liberator," *Agricultural History* 68 (1994): 157–68.

21. Fabian Lange, Alan Olmstead, and Paul Rhode, "The Impact of the Boll Weevil, 1892–1932," *Journal of Economic History* 69, no. 3 (September 2009): 685–718.

22. Higgs's analysis assumes that tenants had no notion that the weevil was coming, or what its arrival would mean, and that the pest influenced one's decision to migrate only after it had destroyed his or her crop. Higgs also claimed that no social factors, including the "brutalities and discriminations heaped upon blacks in the South," were an impetus for the Great Migration since these occurrences had been a factor in black southern life since the Civil War and not created a massive outmigration previously. Higgs, "The Boll Weevil, the Cotton Economy, and Black Migration, 1910–1930," 338, 350, 337; Strickland, "Strange Affair of the Boll Weevil," 165.

23. Lange et al., "Impact of the Boll Weevil," 710–14.

24. There is a vast literature on postbellum southern labor and migration. See, for example, Edward Ayers, *Promise of the New South*; Pete Daniel, *The Shadow of Slavery: Peonage in the South, 1901–1969* (Urbana: University of Illinois Press, 1972); William Cohen, *At Freedom's Edge: Black Mobility and the Southern White Quest for Racial Control, 1861–1915* (Baton Rouge: Louisiana State University Press, 1991); Roger L. Ransom, and Richard Sutch, *One Kind of Freedom: The Economic Consequences of Emancipation*, 2nd ed. (Cambridge: Cambridge University Press, 2001), especially pp. 172–74, 196.

25. Taylor, *American-Mexican Frontier*, 133; Coclanis and Simon, "Exit, Voice, and Loyalty," 205.

26. Taylor, *American-Mexican Frontier*, 133; Coclanis and Simon, "Exit, Voice, and Loyalty," 205.

27. "Hidden transcripts of resistance" comes from James C. Scott, *Domination and the Arts of Resistance: Hidden Transcripts* (New Haven: Yale University Press, 1990); Paul S. Taylor "Corrido de Texas," in Mody C. Boatright, Wilson M. Hudson, and Allen Maxwell, eds., *Texas Folk and Folklore* (Dallas: Southern Methodist University Press, 1954), 157–58.

28. Gates Thomas, "South Texas Negro Work Songs," *Publications of the Texas Folklore Society* 5 (1926): 155.

29. Ibid., 154.

30. Ibid., 175.

31. The chapters that follow continue to chart the movement of the song and will detail how Mississippi bluesmen knew of the pest through song long before the weevil actually made it to the Magnolia State.

32. James Haskins with Kathleen Benson, *Scott Joplin* (Garden City, NY: Doubleday and Co., 1978), 33, 58–62; Edward A. Berlin, *King of Ragtime: Scott Joplin and His Era* (New York: Oxford University Press, 1994), 2–6.

33. Barr, *Black Texans*, 97; Sharpless, *Fertile Ground, Narrow Choices*, 69; Charles Wolfe and Kip Lornell, *The Life and Legend of Leadbelly* (New York: HarperCollins, 1992), 42, 52.

34. Thomas, "South Texas Negro Work Songs," 173.

35. Ibid.

36. "Boll Weevil Song," cited as "'Traditional'" (no specific author), in Hazel Arnett, *I Hear America Singing!: Great Folk Songs From the Revolution to Rock* (New York: Praeger Publishers, 1975), 140–41. One similar version is Eddie Cochran's "Boll Weevil Song," but there are countless others.

37. Thomas, "South Texas Negro Work Songs," 155, 173.

38. Ibid., 174.

39. Ibid.

40. Leadbelly, "Boll Weevil" (audio recording), Library of Congress Recordings (LC 273-A-1 and LC 135-A).

41. Ibid.

42. Willard A. Dickerson et al., eds., *Boll Weevil Eradication in the United States through 1999*, the Cotton Foundation Reference Book Series, no. 6 (Memphis: Cotton Foundation Publisher, 2001), 590, 595–612.

43. United States Bureau of the Census, *Thirteenth Census of the United States* (Washington, D.C.: Government Printing Office, 1912); Dickerson, et al., eds., *Boll Weevil Eradication*, 590, 595–612.

44. Knapp quoted in Bailey, *Seaman A. Knapp*, 169.

45. Scott, *Reluctant Farmer*, 222–25.

CHAPTER 3

1. New Orleans *Picayune*, March 11, 1909; Memphis *Commercial Appeal*, March 3, 1909, 5; Dunbar Rowland, *Mississippi: Comprising Sketches of Counties, Towns, Events, Institutions, and Persons, Arranged in Cyclopedic Form*, vol. 1 (Atlanta: Southern Historical Publishing Association, 1907), 127, 229, 761, 874.

2. "Land and Industrial Agent" to R. V. Taylor, March 29, 1909, John Milliken Parker Papers (hereafter Parker Papers), box 2, folder 27, Southern Historical Collection, University of North Carolina at Chapel Hill.

3. "Farmers Train Drawing Well," *Memphis Commercial Appeal*, March 9, 1909, p. 7; Birmingham *Age-Herald*, March 9, 1909, clipping, in Parker Papers, box 2, folder 25; New Orleans *Picayune*, March 10, 1909, clipping, in Parker Papers, box 2, folder 25.

4. Rowland, *Mississippi*, 803; Land and Industrial Agent to R. V. Taylor, March 29, 1909, Parker Papers, box 2, folder 27.

5. Land and Industrial Agent to R. V. Taylor, March 29, 1909, Parker Papers, box 2, folder 27.

6. Ibid.

7. United States Bureau of the Census, *Twelfth Census of the United States* (Washington, D.C.: Government Printing Office, 1901); United States Bureau of the Census, *Thirteenth*

Census of the United States (Washington, D.C.: Government Printing Office, 1912); United States Bureau of the Census, *Fourteenth Census of the United States* (Washington, D.C.: Government Printing Office, 1922); United States Bureau of the Census, *Fifteenth Census of the United States* (Washington, D.C.: Government Printing Office, 1932); United States Bureau of the Census, *Sixteenth Census of the United States Take in the Year 1940* (Washington, D.C.: Government Printing Office, 1943).

8. I am following the definition of the Yazoo-Mississippi Delta offered by Mikko Saikku, *This Delta, This Land: An Environmental History of the Yazoo-Mississippi Floodplain* (Athens: University of Georgia Press, 2005).

9. Donald H. Bowman, "A History of the Delta Branch Experiment Station, 1904–1985," special bulletin 86–2 (Mississippi State, MS: Mississippi Agricultural and Forestry Experiment Station, August 1986), 65; Cobb, *The Most Southern Place on Earth: The Mississippi Delta and the Roots of Regional Identity* (New York: Oxford University Press, 1992), vii; John M. Barry, *Rising Tide: The Great Mississippi Flood of 1927 and How It Changed America* (New York: Touchstone, 1997), 98; William Lincoln Giles, "Agricultural Revolution, 1890–1970," in Richard Aubrey McLemore, ed., *A History of Mississippi*, vol. 2 (Hattiesburg: University and College Press of Mississippi, 1973): 183; Saikku, *This Delta, This Land*, 115. For a complete discussion of the Delta as frontier, see John Charles Willis, *Forgotten Time: The Yazoo-Mississippi Delta after the Civil War* (Charlottesville: University Press of Virginia, 2000).

10. Cobb, *The Most Southern Place on Earth*, vii; Barry, *Rising Tide*, 98; Giles, "Agricultural Revolution, 1890–1970," 183.

11. Despite twentieth-century planters' belief that they were the first people to inhabit the region, the area had a long human history prior to their arrival. Yet as Mikko Saikku explains, there were long periods without human settlement. Even when homo sapiens was not living there, its presence at other points along the rivers affected the Delta environment. Saikku, *This Delta, This Land*, 2; Harris, *Deep Souths: Delta, Piedmont, and Sea Island Society in the Age of Segregation* (Baltimore: Johns Hopkins University Press, 2001), 42; Steven Stoll, *Larding the Lean Earth: Soil and Society in Nineteenth Century America* (New York: Hill and Wang, 2002), 19; Barry, *Rising Tide*, 97.

12. Joe Rice Dockery, typed transcript of interview by John Jones, December 13, 1979, Mississippi Department of Archives and History.

13. Ibid.; Cobb, *Most Southern Place on Earth*, 79; Harris, *Deep Souths*, 48.

14. Cobb, *Most Southern Place on Earth*, 97.

15. Cobb, *Most Southern Place on Earth*, 81; Harris, *Deep Souths*, 44–45. For a complete discussion of the expansion of railroads in the Delta, see Cobb, *Most Southern Place on Earth*, 80; Brandfon, *Cotton Kingdom of the New South*, especially chapter 4; and Willis, *Forgotten Time*.

16. Brandfon, *Cotton Kingdom of the New South*, 91–92.

17. Anne O'Hare McCormick, as quoted in David L. Cohn, *The Mississippi Delta and the World: The Memoirs of David L. Cohn*, James C. Cobb, ed. (Baton Rouge: Louisiana State University Press), 61; David L. Cohn, *Where I Was Born and Raised* (Boston: Houghton Mifflin Company, 1948), 41 (block quote).

18. "Labor Famine in Georgia," *Progressive Farmer* (Raleigh edition), January 16, 1900, 3;

Robert L. Brandfon, "The End of Immigration to the Cotton Fields," *Mississippi Valley Historical Review*, 50, no. 4 (March 1964): 592.

19. E. A. Boeger and E. A. Goldenweiser, "A Study of the Tenant Systems of Farming in the Yazoo-Mississippi Delta," United States Department of Agriculture, bulletin no. 337, January 13, 1916 (Washington, D.C: Government Printing Office, 1916), 3; Brandfon, *Cotton Kingdom of the New South*, 56.

20. Writing on the evolution of the crop lien system and the evolution of the sharecropping system is well known. See Harold Woodman, *King Cotton and His Retainers: Financing & Marketing the Cotton Crop of the South, 1800 –1925* (Lexington: University of Kentucky Press, 1968); Woodman, *New South, New Law: The Legal Foundations of Credit and Labor Relations in the Postbellum Agricultural South* (Baton Rouge: Louisiana State University Press, 1995); Ransom and Sutch, *One Kind of Freedom*, Gavin Wright, *Old South, New South: Revolutions in the Southern Economy since the Civil War* (New York: Basic Books, 1986).

21. Walter Sillers, Sr., to P. M. Burrill, December 31, 1907, folder 19A, Walter Sillers, Sr. Papers, Delta State University Archives (hereafter Sillers Papers).

22. Vicksburg *Herald* article reprinted in Gloster (Mississippi) *Herald*, October 23, 1903, 4; "Fund to Fight Boll Weevil," *Memphis Commercial Appeal*, January 6, 1909, 3; "The Cotton Crop is Doing Well," Greenwood *Commonwealth*, August 7, 1908; Greenwood *Commonwealth*, July 9, 1909; Greenwood *Commonwealth*, October 30, 1908.

23. "The Hot Weather Burns Boll Weevils," Greenwood *Commonwealth*, August 27, 1909; "Supposed Boll Weevils," Woodville *Republican*, reprinted in Greenwood *Commonwealth*, February 19, 1909; Vicksburg *Herald* article reprinted in Gloster (Mississippi) *Herald*, October 23, 1903, 4.

The Vicksburg *Herald*'s insistence that farmers not import seeds to the Delta from the infested territory echoed a cry going out across the as-yet-infected regions of the South. The Georgia Agricultural Commissioner had in fact enacted a ban against Texas and Louisiana "cotton, cotton seed, hulls, corn, hay or other farm products" in which weevils might hide away.

24. Theodore Roosevelt, "Fourth Annual Message to Congress," December 6, 1904; W. D. Hunter, "Some Recent Studies of the Mexican Cotton Boll Weevil," *United States Department of Agriculture Yearbook of Agriculture for 1906* (Washington, D.C.: Government Printing Office, 1907), 313.

25. "Insects and Diseases Liable to be Introduced into Mississippi," Mississippi Agricultural Extension Service Bulletin, no. 96 (February 1906). The department finally made mention of the pest that year, noting that the bug, along with six other insects, was "liable to be introduced into Mississippi" during that year's season. "It is probable that every one in the State has heard of the Mexican Cotton Boll Weevil," the bulletin admitted, but noted that few Mississippians could recognize it. The pamphlet included three photographs of weevils in various stages of development and a detailed physical description. The department admitted that the pest "will eventually reach our cotton fields," and vaguely encouraged farmers to "do all in our power to retard its coming as long as we can."

26. R. W. Harned, "Boll Weevil in Mississippi, 1909," Mississippi Agricultural Extension Service Bulletin, no. 139, March 1910; Giles, "Agricultural Revolution, 1890–1970," 191;

W. L. Hutchinson, "Cotton Culture in Mississippi (In areas infested with the Mexican Cotton Boll Weevil)," Mississippi Agricultural Extension Service Bulletin, no. 117, December 1908.

27. Alfred Holt Stone, "The Negro in the Yazoo-Mississippi Delta," *Publications of the American Economic Association*, 3rd series, vol. 3, no. 1 (February 1902): 235–36. Not surprisingly, Stone was a good friend of the racist environmental determinist historian U. B. Phillips, who actually visited Stone's plantation prior to writing *American Negro Slavery*. John David Smith, "Alfred Holt Stone: Mississippi Planter and Archivist/Historian of Slavery," *Journal of Mississippi History* 45 (1983): 262–63. Recent work on Stone concentrates on this dual obsession with race and science. See James G. Hollandsworth, Jr., *Portrait of a Scientific Racist: Alfred Holt Stone of Mississippi* (Baton Rouge: Louisiana State University Press, 2008).

28. Stone, "The Negro in the Yazoo-Mississippi Delta" 239, 240.

29. Greenville *Times*, December 19, 1909; emphasis in original; "Advice to Negro Tenants," Greenwood *Commonwealth*, November 6, 1908. See also Lange et al., "Impact of the Boll Weevil."

30. *Laws of the State of Mississippi Passed at a Regular Session of the Mississippi Legislature* (Nashville: Brandon Printing Company, 1904), 115; *Journal of the House of Representatives of the State of Mississippi at a Regular Session Thereof* (Nashville: Brandon Printing Company, 1904), 321, 444; *Journal of the Senate of the State of Mississippi of a Regulars Session Thereof* (Nashville: Brandon Printing Company, 1904); Bowman, "A History of the Delta Branch Experiment Station, 1904–1985," 2–4.

31. Eugene B. Ferris, "Early Recollections of the Mississippi A. & M. College," unpublished manuscript, Ferris Family Papers, Mississippi Department of Archives and History, box 7, folder 111, 3; hereafter Ferris Papers; Bowman, "A History of the Delta Branch Experiment Station, 1904–1985," 2–4.

32. Eugene B. Ferris offers a critical firsthand view of early extension work with Mississippi farmers in "Early Recollections of the Mississippi A. & M. College," Ferris Papers, 2–3; J. W. Fox, "Report of the Work at the Delta Station for 1907–8," Mississippi Agricultural Experiment Station (MAES) Bulletin, no. 119, March 1909 [date on first page is misprinted as 1907], 2; Giles, "Agricultural Revolution, 1890–1970," 191–92. Delta newspapers also recorded their support of the early work of John Fox, the director of the Stoneville station. See Greenville *Times*, July 24, 1909, and July 17, 1909. Delta planters like Walter Sillers commonly wrote to Fox and other scientists asking advice about when to apply fertilizers, when to plant crops, and what seed to select. See for example Walter Sillers, Sr., to J. W. Fox, May 10, 1911, in Sillers Papers, folder 21. There is a growing body of work on the racism inherent in southern agricultural extension. See Daniel, *Breaking the Land*; Daniel, *Shadow of Slavery: Peonage in the South, 1901–1969* (Urbana: University of Illinois Press, 1972); Reid, *Reaping a Greater Harvest*.

33. J. W. Fox, "Report of the Work at the Delta Station for 1907–8," 6; LeRoy Percy to Bolton Smith, December 31, 1908, in Percy Papers, box 4, folder 5. For larger context of fight within USDA over approaches to the boll weevil, see Giesen, "The South's Greatest Enemy?," chapters 1 and 2; Helms, "Just Lookin' For a Home," chapter 2; Hae-Gyung Geong, "Exerting Control: Biology and Bureaucracy in the Development of American Entomology, 1870–1930" (Ph.D. dissertation, University of Wisconsin-Madison, 1999), chapters 1 and 2.

34. Fox, "Report of the Work at the Delta Station for 1907–8," 6 (italics mine). A year

later, Fox and his colleagues expanded the advice for battling the boll weevil to include the USDA's cultural method, which advised planting a quickly maturing seed early in the season and picking it as early as possible, as well as new techniques for row spacing, fertilizer use, and field crop rotation. R. W. Harned, "Boll Weevil in Mississippi, 1909," MAES Bulletin, no. 139, March 1910.

35. LeRoy Percy to John G. Jones, December 3, 1908, in Percy Papers, box 4, folder 5.

36. David L. Cohn, *The Mississippi Delta and the World: The Memoirs of David L. Cohn,* James C. Cobb, ed. (Baton Rouge: Louisiana State University Press), 65.

37. "Lectures to Farmers," Greenwood *Commonwealth*, August 14, 1908; "Literature on the Boll Weevil," Greenwood *Commonwealth*, October 30, 1908.

38. "Literature on the Boll Weevil," Greenwood *Commonwealth*, October 30, 1908.

39. Alfred H. Stone and Julian H. Fort, "The Truth About the Boll Weevil" (Greenville, Mississippi: First National Bank, 1911), 3.

40. Ibid., 3.

41. Ibid, 7, 9, 21.

42. Ibid., 6.

43. Ibid, 30, 31.

44. Ibid, 33.

45. LeRoy Percy to John G. Jones, December 3, 1908, in Percy Papers, box 4, folder 5.

46. John G. Jones of the Illinois Central Railroad wrote to LeRoy Percy in 1908, also fearful of the spread of pessimistic news of the boll weevil by his railroad's own educational train, the Boll Weevil Special. He wrote to Percy that he "cannot concur in the opinion of the government experts that our country is 'going to the bad' on account of the Boll weevil [*sic*], and am anxious to secure from you and other representative planters in the delta an expression of your sentiments to be embodied in a pamphlet to be distributed on this trip. It is my opinion that we should take as optimistic a view as possible of this matter, at the same time appreciating the necessity of taking such steps as may be necessary to guard against the advent of the boll weevil. As I feel you you [*sic*] have given this matter considerable thought, I hope you will kindly favor me with an expression of your views." John G. Jones to LeRoy Percy, December 1, 1908, in Percy Papers, box 4, folder 5.

47. Greenville (Mississippi) *Daily Times*, December 19, 1909, 4; Giles, "Agricultural Revolution, 1890–1970," 198.

48. Cohn, *Where I Was Born and Raised*, 41–42.

49. Stewart, "If John Muir Had Been an Agrarian," 150.

50. LeRoy Percy to Johanna Reiser, October 19, 1908, in Percy Papers, box 21, folder 2. Harris, *Deep Souths*, 128–31, details some of this exchange of letters.

51. LeRoy Percy to Johanna Reiser, October 19, 1908, in Percy Papers, box 21, folder 2.

52. Ibid.

53. LeRoy Percy to Johanna Reiser, December 31, 1908, in Percy Papers, box 21, folder 2.

54. Ibid.

55. Ibid; LeRoy Percy to Sophie Reiser, January 26, 1909, in Percy Papers, box 21, folder 3.

56. LeRoy Percy to Johanna Reiser, November 24, 1909, in Percy Papers, box 21, folder 4; LeRoy Percy to Johanna Reiser, June 1, 1910, in Percy Papers, box 21, folder 5.

57. William Percy to Will Howe, November 7, 1910, in Percy Papers, box 21, folder 6.

58. William A. Percy to Johanna Reiser, July 15, 1910, in Percy Papers, box 21, folder 5.

59. William Percy to Johanna Reiser, November 10, 1910, in Percy Papers, box 21, folder 6; Johanna Reiser to William Percy, October 14, 1910, in Percy Papers, box 21, folder 5.

60. Johanna Reiser to William Percy, October 14, 1910, in Percy Papers, box 21, folder 5.

61. Ibid.

62. William Percy to Johanna Reiser, December 31, 1910, in Percy Papers, box 21, folder 7.

63. William Percy to Johanna Reiser, March 14, 1911, in Percy Papers, box 21, folder 7.

CHAPTER 4

1. Francis Getze oral history transcript, Delta and Pine Land Company Records, Special Collections, Mississippi State University (hereafter DPLC Papers), series 16: Oral History, folder 17; Hobson, "Delta & Pine Land Co.," January 22, 1937, in DPLC Papers, series 9: "The History of the Delta and Pine Land Company," box 27, folder 7.

2. Walter Sillers, Sr., to P. M. Burrill, March 22, 1909, in Sillers Papers, folder 19A.

3. W. B. Mercier and H. E. Savely, *The Knapp Method of Growing Cotton* (Garden City: Doubleday, Page & Company, 1913), 116.

4. Fabian Lange, Alan Olmstead, and Paul Rhode found that farmers seeking to get one last big crop drove up the cotton supply ahead of the boll weevil's migration. Lange et al, "Impact of the Boll Weevil," 704–5.

5. Cotton prices climbed steadily in the second half of the nineteenth and early twentieth centuries. The agricultural South as a whole was still rebuilding its lands from the damage of the Civil War. And though many foreign mills, cut off from American cotton in wartime, had turned to the growing markets of India and Egypt, there was still not enough competition to sway southerners from putting more land into cotton production. Not only was the supply limited, but British mills faced expanding competition from other countries in Europe, as well as Japan and the United States, for the raw material. As more and more spindles were put into production around the world, the demand for and price of cotton at the turn of the century soared. Despite the region's growth, at the turn of the century the Delta could only claim that 30 percent of its land was in production. The role of cotton prices in the boll weevil's history is explored in detail in subsequent chapters. Cobb, *Most Southern Place on Earth*, 100; Brandfon, *Cotton Kingdom of the New South*, 117–18.

6. Brandfon, *Cotton Kingdom of the New South*, 127–28; Lawrence J. Nelson, *King Cotton's Advocate: Oscar G. Johnston and the New Deal* (Knoxville: University of Tennessee Press, 1999), 24.

7. Nelson, *King Cotton's Advocate*, 25.

8. Ibid, 24.

9. Hobson, "Delta & Pine Land Co.," in DPLC Papers, series 9: "The History of the Delta and Pine Land Company," box 27, folder 6; Brandfon, *Cotton Kingdom of the New South*, 119; "English Negotiate for Land in Delta," Memphis *Commercial Appeal*, April 1, 1911, 4.

10. Hobson, "Delta & Pine Land Co.," in DPLC Papers, series 9: "The History of the Delta and Pine Land Company," box 27, folder 7. Before the deal could be finalized, however, Salsbury and the Fine Spinners had to find a way around an old Mississippi Constitution rule that forbade non–United States citizens from owning land in the state. As a result, Salsbury

put together a complicated deal involving a holding company, the division of the lands into different plantations, and the purchase of a separate business charter. As a result, the Fine Spinners ended up owning the Mississippi Delta Planting Company, which leased the land in two pieces, the Triumph Plantation Company and the Lake Vista Plantation Company. Later, the group purchased the charter to the Delta and Pine Land Company, which had been granted prior to the 1890 Mississippi Constitution, which had outlawed foreign landownership, allowing the Fine Spinners to grandfather their way past more recent regulations on businesses, giving the "new" corporation immense power to organize the land and its holding in a number of ways. The Mississippi Delta Planting Company purchased the DPLC charter in 1919. Since there were no major changes in ownership that came with the charter purchase, for simplicity I refer to the company as DPLC even prior to the 1919 official moniker change. Harris, *Deep Souths*, 122; Nelson, *King Cotton's Advocate*, 24.

11. "Three Million Is Invested in Delta," Memphis *Commercial Appeal*, May 18, 1911, 1.

12. Harris, *Deep Souths*, 122; Nelson, *King Cotton's Advocate*, 24.

13. Nelson, *King Cotton's Advocate*, 24; Mrs. Early C. Ewing, "The Delta and Pine Land Company," in *History of Bolivar County, Mississippi*, ed. Wirt A. Williams (Spartanburg, SC: Reprint Company, 1976), 250. In 1910, Fox had been appointed director of the MAES and was based in Starkville, though he continued to oversee the research and education projects in the Delta. Giles, "Agricultural Revolution, 1890–1970," 192.

14. J. W. Fox, "Report of Work at the Delta Station for 1906," Mississippi Agricultural Extension Service Bulletin, no. 106, January 1907; New Orleans *Picayune*, March 11, 1909; Early C. Ewing, Jr., transcript of interview by Roberta Miller, June 5, 1978, William Alexander Percy Public Library, Greenville, Mississippi (hereafter Ewing Transcript). See also Eugene B. Ferris, "Early Recollections of the Mississippi A. & M. College," unpublished manuscript, Ferris Papers, box 7, folder 111, p. 8.

15. J. W. Fox, "Report of the Work at the Delta Station for 1907–8," Mississippi Agricultural Extension Service Bulletin, no. 119, March 1909 [date on first page is misprinted as 1907].

16. Nelson, *King Cotton's Advocate*, 28.

17. Robert W. Harrison, *Alluvial Empire* (Little Rock: Pioneer Press, 1961), 129; Mrs. Early C. Ewing, "The Delta and Pine Land Company," 250; Ewing, "History of the Delta and Pine Land Company, 1911–1967," n.d., DPLC Papers, 8. Floods and levee construction have their own sordid history in the Delta, and many of the issues pertinent to the advancing boll weevil were being sorted out by the region's people before and after the cotton pest's arrival. Floods were an environmental force that would also summon federal workers to the region.

18. Nelson, *King Cotton's Advocate*, 29.

19. A. C. Wild to G. Patterson, December 22, 1965, in Early C. Ewing, "History of the Delta and Pine Land Company, 1911–1967," n.d., DPLC Papers, 7.

20. Brandfon, *Cotton Kingdom of the New South*, 129; Vernon Bellhouse, "Delta & Pine Land Company," September 30, 1948, in DPLC Papers, series 9: "The History of the Delta and Pine Land Company," box 27.

21. Ewing Transcript, p. 3.

22. Finding Aid, DPLC Papers, n.d.; Ewing Transcript; Ewing, "History of the Delta and Pine Land Company, 1911–1967," n.d., DPLC Papers, 12.

23. Early C. Ewing Sr., oral history transcript, DPLC Papers, series 16: Oral History, folder 16; Ewing, "History of the Delta and Pine Land Company, 1911–1967," n.d., DPLC Papers, 12–14.

24. Annual Report of the Experimental Department, Mississippi Delta Planting Company, April 7, 1916, in DPLC Papers, series 5: Early C. Ewing, Sr. Papers, box 32, folder v-a1.

25. Ewing, "History of the Delta and Pine Land Company, 1911–1967," n.d., DPLC Papers, 13–15; G. B. Walker, "Report of Work at the Delta Branch Experiment Station For 1911," Mississippi Agricultural Experiment Station Bulletin, no. 157, February 1912.

26. Annual Report of the Experimental Department, Mississippi Delta Planting Company, April 7, 1916, in DPLC Papers, series 5: Early C. Ewing, Sr. Papers, box 32, folder v-a1. See also "Recent Cotton Experiments," Mississippi Agricultural Experiment Station Bulletin, no. 155, December 1911; and Early C. Ewing, Jr., "The Cotton Research Program of the Delta & Pine Land Co.," September 27, 1960, in DPLC Papers, series 9: "The History of the Delta and Pine Land Company," box 27.

27. Early C. Ewing, Jr., oral history transcript, DPLC Papers, series 16: Oral History, folder 15.

28. Mrs. Early C. Ewing, "The Delta and Pine Land Company," 251.

29. Helms, "Just Lookin' for a Home: The Cotton Boll Weevil and the South," 95; Edmund Russell, *War and Nature: Fighting Humans and Insects with Chemicals from World War I to* Silent Spring (Cambridge: Cambridge University Press, 2001), 6; Shirley A. Briggs and Staff of Rachel Carson Council, *Basic Guide to Pesticides: Their Characteristics and Hazards* (Washington: Hemisphere Publishing Corporation, 1992), 60.

30. Brodie S. Crump, transcript of interview by Lelia Clark Wynn, June 1, 1977, William Alexander Percy Public Library, Greenville, Mississippi, p. 18; Elmer Johnson and B. R. Coad, "Dusting Machinery for Cotton Boll Weevil Control," U.S. Department of Agriculture Farmers Bulletin, no. 1098 (January 1920), 5.

31. Greenville *Daily Times*, January 9, 1910; Ewing, "History of the Delta and Pine Land Company, 1911–1967," n.d., DPLC Papers, 10. Officials with the USDA in Washington reacted skeptically to Newell's findings. The department understood the difficulties with application and continued to put the focus of farmer education on the cultural method rather than on poison. The USDA was eventually swayed from this stance first by the entry of other government agencies into the boll weevil fight and later by the development of more economical and efficient insecticide sprayers. The former came as the result of World War I. When the war was over, many of the strategies developed by the military to fight human enemies were put to use against the boll weevil and other insects. The Chemical Warfare Service spearheaded tests of chemicals against the boll weevil but made little ground. As historian Edmund Russell has explained, the War Department played a major role in the battle against insect enemies on the home front. He also notes that the poisonous gases used in battle during World War I had no effect on the boll weevil. Russell, *War and Nature*, 64–65.

32. Early C. Ewing, "Cotton in the Mid-South: Highlights of Agricultural History in the Central Belt," in DPLC Papers, n.d.; Ewing, "History of the Delta and Pine Land Company, 1911–1967," n.d., DPLC Papers, 10, 11.

33. Mrs. Early C. Ewing, "The Delta and Pine Land Company," 250; R. M. (Dick) Holman, oral history transcript, DPLC Papers, series 16: Oral History, folder 27.

34. Ernest Haywood, oral history transcript, DPLC Papers, series 16: Oral History, folder 22; Ewing, "History of the Delta and Pine Land Company, 1911–1967," n.d., DPLC Papers, 12; Ewing Transcript, p. 4.

35. W. David Lewis and Wesley Phillips Newton, *Delta: The History of an Airline* (Athens: University of Georgia Press, 1979), 10–12. The first successful crop dusting experiments were made in Ohio. Concurrent to Coad's work in Louisiana was a large-scale dusting experiment on fruit trees with planes based in Macon, Georgia. Pete Daniel, *Lost Revolutions: The South in the 1950s* (Chapel Hill: University of North Carolina Press, 2000), 63.

36. Lewis and Newton, *Delta*, 12–13; Daniel, *Lost Revolutions*, 63; B. R. Coad to George Patterson, January 5, 1966, in DPLC Papers, series 9.

37. B. R. Coad to George Patterson, January 5, 1966, in DPLC Papers, series 9.

38. Lewis and Newton, *Delta*, 10–12, 39; Pete Daniel, *Lost Revolutions*, 63; Daniel, *Toxic Drift: Pesticides and Health in the Post–World War II South* (Baton Rouge: Louisiana State University Press, 2005).

39. B. R. Coad, "Dusting Cotton From Airplanes," USDA Bulletin no. 1204 (GPO 1924), 23–24, 36.

40. R. M. (Dick) Holman, oral history transcript, DPLC Papers, series 16: Oral History, folder 27; Early Ewing Jr., oral history transcript, DPLC Papers, series 16: Oral History, folder 15; Daniel, *Toxic Drift*.

41. Ewing Transcript, pp. 7, 21.

42. Untitled manuscript, October 28, 1943, in DPLC Papers, series 9: "The History of the Delta and Pine Land Company," box 27; Nelson, *King Cotton's Advocate*, 97. See also James C. Giesen, "'The Truth About the Boll Weevil': The Nature of Planter Power in the Mississippi Delta," *Environmental History* 14, no. 4 (October 2009).

43. Cornelius Bostick, oral history transcript, DPLC Papers, series 16: Oral History, folder 5; R. M. (Dick) Holman, oral history transcript, DPLC Papers, series 16: Oral History, folder 27.

44. Boeger and Goldenweiser, "A Study of the Tenant Systems of Farming in the Yazoo-Mississippi Delta," 15; John Charles Willis, "On the New South Frontier: Life in the Yazoo-Mississippi Delta, 1865–1920," (Ph.D. Dissertation, University of Virginia, 1991), 356–57.

45. Minor S. Gray, "A Short History of Delta and Pine Land Company," unpublished manuscript, n.d., in DPLC Papers, series 9, box 27, p. 5; Ewing Transcript, p. 26.

46. Minor S. Gray, "A Short History of Delta and Pine Land Company," unpublished manuscript, n.d., in DPLC Papers, series 9: "The History of the Delta and Pine Land Company," box 27, p. 6.

47. For one traditional account of pervasive chicanery and peonage in sharecropping, see Daniel, *Shadow of Slavery*.

48. "Biggest Cotton Plantation," *Fortune* 15, no. 3 (March 1937), 125, 126.

49. R. M. (Dick) Holman, oral history transcript, DPLC Records, series 16: Oral History, folder 27.

50. Early C. Ewing, Jr., transcript of interview by Roberta Miller, June 5 1978, William Alexander Percy Public Library, Greenville, MS. Ewing Transcript, p. 22.

51. Henry C. Speir interview (Jackson, Mississippi, May 18, 1968) by Gayle Dean Wardlow

on *Screamin' and Hollerin' the Blues: The Worlds of Charley Patton*, Revenant Album no. 212 (seven-compact disc boxed set, 2001); James C. Cobb, "The Blues Is a Lowdown Shakin' Chill," in *Redefining Southern Culture: Mind and Identity in the Modern South* (Athens: University of Georgia Press, 1999), 103.

52. David Evans, "Charley Patton: The Conscience of the Delta," essay in liner notes of *Screamin' and Hollerin' the Blues: The Worlds of Charley Patton*, 10.

53. Ibid, 12, 13, 15.

54. Stephen Calt and Gayle Wardlow, *King of the Delta Blues: The Life and Music of Charlie Patton* (Newton, New Jersey: Rock Chapel Press, 1988), 11–13, 280; Henry C. Speir interview.

55. John Fahey, *Charley Patton* (London: November Books, 1970), 66; Booker Miller, interviewed by Gayle Dean Wardlow (1968), on *Screamin' and Hollerin' the Blues: The Worlds of Charley Patton*.

56. Tom Piazza, *Blues and Trouble: Five Stories* (New York: St. Martins Press, 1996), 190.

57. Charley Patton, "Mississippi Bo Weavil Blues," Paramount 12805 (1929).

58. The recording itself, as well as the lyrics, are draped in ambiguity and secrecy. Its first recorded incarnation—that 1929 Paramount 78-RPM record—was not initially even attributed to Patton. The company marketed it as the work of "The Masked Marvel." Advertisements featured a sketch of a man in a black jacket and bow tie, wearing an eye mask apparently to protect his identity. The accompanying caption asked "Guess Who He Is?" Paramount released other recordings of the song at the same time, which they correctly attributed to Patton; the creation of his masked alter ego was simply a marketing ploy. To add even more confusion to the mix, Paramount released a small quantity of the recordings correctly attributing the song to Patton. Nevertheless, it adds to the mystique of a song that is confusing enough on its own. Liner notes of *Screamin' and Hollerin' the Blues: The Worlds of Charley Patton*, appendix 1, Thematic Catalogue of the Recorded Music of Charley Patton, 94.

59. Harris, *Deep Souths*, 244.

60. I have addressed historians' miscalculation of the weevil's affect on migration in chapter 2.

61. Jere B. Nash, transcript of interview by Roberta Miller, May 31, 1977, William Alexander Percy Public Library, Greenville, Mississippi.

62. Giles, "Agricultural Revolution, 1890–1970," 181.

63. Between 1911 and 1927, in fact, the company paid its investors a dividend only once. Then the 1927 Mississippi River flood, the worst of the twentieth century, took more than three thousand acres out of production and cost the plantation an estimated $500,000. During the Depression the company began to consistently turn a profit, thanks in part to DPLC's arrangement with President Roosevelt's Agricultural Adjustment Administration. "Biggest Cotton Plantation," *Fortune* 15, no. 3 (March 1937): 130–31.

CHAPTER 5

1. "Why the Deadly Boll Weevil, Bringing Revolution With Him, Is Called the 'Prosperity Bug,'" *New York Times*, January 9, 1910, part 5, p. 13.

2. Ibid.

3. Ibid.

4. Part of the explanation for Texas's expansion of cotton acreage is that the boll weevil did not thrive in the dry fields of central and western Texas. As a result, farmers there expanded the cotton-growing belt west, towards these weevil-resistant climates.

5. From 1909, the year the weevil entered the Delta, until 1919, planters increased their cotton acreage from 724,815 to 1.18 million. Though the boll weevil is estimated to have destroyed as much as 20 percent of Mississippi's cotton statewide during these ten years, the Delta increased its production by 64 percent. *Thirteenth Census of the United States Taken in the Year 1910*, vol. 6: Agriculture, 1909 and 1910 (Washington, D.C.: Government Printing Office, 1913); *Fourteenth Census of the United States Taken in the Year 1920*, vol., part 2: Agriculture (Washington, D.C.: Government Printing Office, 1922); Dickerson et al., eds., *Boll Weevil Eradication in the United States through 1999*, 590.

6. I define the region of southeastern Alabama as the counties of Barbour, Bullock, Coffee, Covington, Crenshaw, Dale, Geneva, Henry, Houston, Lee, Macon, Pike, and Russell. Houston County was created in 1903 from parts of Henry, Dale, and Geneva Counties. See Fig. 10.

7. Soil maps of the Deep South indicate three major bands of land types that run from west to east through Alabama. Within each major region, however, dozens of local variations of soil, topography, and geology exist. The skinny southernmost region is a level coastal plain that runs from the Florida border about sixty miles to the north. The soils of this region, which contains Covington, Geneva, and Houston Counties, are rich in nutrients, relatively flat, and productive. North of this section, making up more than its share of the lower half of the state, is the hilly coastal plain sometimes called the Wiregrass. Like its southern neighbor, the land has mixed sandy loam soils where the land is flat, but many hills that roll through the countryside breaking up potential farmland. Comprising most of six counties, a 1953 state study called the soils of this region "badly mixed in character," and "very subject to erosion." A third band runs to the north of these sandy hills and is comprised of flat, dark soils. Most of Macon, Lee, Bullock, and Russell Counties are within this Black Belt region, though by the early twentieth century the lands were not all in equal condition. In Lee County, the "grayish brown to red soils" were the state's most severely eroded. At midcentury the state was willing to admit that Lee's land had washed away to the point where "many areas [have] gone out of crop use." Mary Elizabeth Hines, "Death at the Hands of Persons Unknown: The Geography of Lynching in the Deep South, 1882–1910" (Ph.D. dissertation, Louisiana State University, 1992); Alabama Department of Agriculture, "Soil Map of Alabama" (Montgomery: Alabama Department of Agriculture and Industries, 1953). See also Ransom and Sutch, *One Kind of Freedom*, 276–79.

8. United States Department of Agriculture, Soil Conservation Service, *Soil Survey of Coffee County, Alabama* (Washington: Government Printing Office, March 1979).

9. *Twelfth Census of the United States: 1900* (Washington, D.C.: Government Printing Office, 1902).

10. Charles Grayson Summersell, *Alabama History for Schools*, 5th ed. (Montgomery: Viewpoint Publications, 1975), 439.

11. *Thirteenth Census of the United States Taken in the Year 1910*, vol. 1: *Population*; vol. 6: *Agriculture, 1909 and 1910* (Washington, D.C.: Government Printing Office, 1913).

12. Norwood Allen Kerr, *A History of the Alabama Agricultural Experiment Station, 1883–1983* (Auburn: Alabama Agricultural Experiment Station and Auburn University, 1985), 29, 34–35.

13. Louis R. Harlan, ed., *The Booker T. Washington Papers*, vol. 3 (Urbana: University of Illinois Press, 1974), 583–87.

14. Mark Hersey explores Carver's broad environmental outlook in "'My Work Is That of Conservation': The Environmental Vision of George Washington Carver" (Ph.D. dissertation, University of Kansas, 2006). See also Hersey, "Hints and Suggestions to Farmers: George Washington Carver and Rural Conservation in the South," *Environmental History* 11 (April 2006), 239–68.

15. Robert J. Norrell, *Reaping the Whirlwind: The Civil Rights Movement in Tuskegee*, reprint ed. (Chapel Hill: University of North Carolina Press, 1998), 22; Linda O. Hines, "White Mythology and Black Duality: George W. Carver's Response to Racism and the Radical Left," *Journal of Negro History* 62, no. 2 (April 1977): 135; Lewis W. Jones, "The South's First Negro Farm Agent," *Journal of Negro Education* 22, no. 1 (Winter, 1953): 38–39; Lu Ann Jones, *Mama Learned Us to Work: Farm Women in the New South* (Chapel Hill: University of North Carolina Press, 2002), 16. See also Allen Jones, "Improving Rural Life for Blacks: The Tuskegee Negro Farmers' Conference, 1892–1915," *Agricultural History* 65, no. 2 (1991): 105–14.

16. United States Department of Agriculture, "The Man Who Works with His Hands: Address of President Roosevelt at the Semi-Centennial Celebration of the Founding of Agricultural Colleges in the United States, at Lansing, Michigan, May 31, 1907," circular no. 24 (July 1, 1907), 10.

17. James Wilson, "Report of the Secretary," *Yearbook of the United States Department of Agriculture, 1910* (Washington: Government Printing Office, 1911), 81.

18. Ibid., italics mine. Part of the confusion and failure of previous diversification efforts, farm agents argued, was the limits of federal policy. In its 1903 bill to fund Knapp's demonstration system, Congress had authorized agents to work only with farmers within the area already invaded by the boll weevil. This precluded the extension service from preparing farmers for the pest. In January 1909, Knapp pressed southern Congressmen for $250,000 to expand the work of his agents into the rest of the South. Senator Murphy J. Foster of Louisiana introduced a resolution reflecting Knapp's request, but it died in the agriculture committee. Other senators tried to introduce similar appropriations, but Congress refused to act. Despite the rhetoric of the president and USDA, federal legislators were hesitant to support education bills that were perceived as helping one area of the country more than others. "Fund to Fight Boll Weevil" *Memphis Commercial Appeal*, January 6, 1909, p. 3; Congress, Senate, Senator Foster of Louisiana, 60th Cong., 2nd sess., *Congressional Record* (January 9, 1909): 688. Early in 1909, Senators F. M. Simmons of North Carolina and Anselm J. McLaurin of Mississippi introduced separate bills to further funding for the boll weevil fight. Congress, Senate, Senator Simmons of North Carolina, 60th Cong., 2nd sess., *Congressional Record* (February 9, 1909): 2079; Congress, Senate, Senator McLaurin of Mississippi, 60th Cong., 2nd sess., *Congressional Record* (February 17, 1909): 2561.

19. Knapp quoted in Martin, *Demonstration Work*, 30, 31.

20. The problems with diversification and southerners' dependence on cotton has been explored by a host of scholars. See Charles S. Aiken, *The Cotton Plantation South Since the*

Civil War; Ransom and Sutch, *One Kind of Freedom*; Daniel, *Breaking the Land*; Kirby, *Rural Worlds Lost*; Fite, *Cotton Fields No More*; Wright, *Old South, New South*.

21. One small booklet in Duggar's papers, for instance, promised that "there is not even a remote probability that the boll weevil will ever be exterminated . . . it will eventually be distributed all over the cotton belt." Joseph Hillman, *The Cultivation of Cotton: A Short Treatise Specially Bearing on Fertilization and the Control of the Ravages of the Boll Weevil* (New York: William S. Myers, 1905), in Duggar Family Papers, box 7, folder 79, Auburn University Special Collections; W. E. Hinds, "Heading Off Boll Weevil Panic," Alabama Agricultural Experiment Station Bulletin, no. 159 (December 1911), 228; USDA, Bureau of Plant Industry, Farmers' Cooperative Demonstration Work, Annual Report of Progress, Alabama, 1910, in Alabama Cooperative Extension Service Records, Auburn University (hereafter ACES Records), box 355.

22. USDA, Bureau of Plant Industry, Farmers' Cooperative Demonstration Work, "Annual Report of Progress," Alabama, 1910, in ACES Records, box 355.

23. Twenty-third Annual Report of the Agricultural Experiment Station of the Alabama Polytechnic Institute (January 31, 1911), 19–20; "Duggar Investigates Alabama Cotton Pests," newspaper clipping [no paper name], September 1, 1911, in Warren E. Hinds Papers, box 1, folder 6, Auburn University Special Collections (hereafter Hinds Papers).

24. Twenty-second Annual Report of the Agricultural Experiment Station of the Alabama Polytechnic Institute (January 31, 1910), 18; Warren E Hinds to Cotton Demonstration Agents, July 27, 1911, in Hinds Papers, box 1, folder 1; B. L. Moss to All Alabama Agents, July 19, 1912, in Hinds Papers, box 1, folder 2.

25. W. E. Hinds, "Heading Off Boll Weevil Panic," Alabama Agricultural Experiment Station Bulletin, no. 159 (December 1911), 228.

26. Hinds, "Heading Off Boll Weevil Panic," 234–36.

27. Kerr, *A History of the Alabama Agricultural Experiment Station*, 35–37.

28. The best treatment of the precarious position that black extension agents—a group that was educated and middle-class—is offered by Jeanne Whayne and Debra Reid. See Whayne, "'I Have Been through Fire': Black Agricultural Extension Agents and the Politics of Negotiation," in Douglas Hurt, ed., *African American Life in the Rural South, 1900–1950* (Columbia: University of Missouri Press, 2003); Reid, *Reaping a Greater Harvest*.

29. George W. Carver, "Fertilizer Experiments on Cotton," Tuskegee Experiment Station Bulletin, no. 3, 1899; George W. Carver, "Cow Peas," Tuskegee Experiment Station Bulletin, no. 5, 1903; George W. Carver, "How to Build Up Worn Out Soils," Tuskegee Experiment Station Bulletin, no. 6, 1905. See also Hersey, "Hints and Suggestions to Farmers," 244–46.

30. Jones, "The South's First Negro Farm Agent," 38–39. In 1897, the Alabama legislature had approved funding for an experiment station on the Tuskegee campus, but offered no support for public education. Allen W. Jones, "The Role of Tuskegee Institute in the Education of Black Farmers," *Journal of Negro History* 60, no. 2 (April 1975): 257.

31. B. D. Mayberry, "The Tuskegee Movable School: A Unique Contribution to National and International Agriculture and Rural Development," *Agricultural History* 65, no. 2 (1991): 87.

32. Ibid.

33. Ibid.

34. Booker T. Washington to Seaman A. Knapp, November 9, 1906, in Thomas Monroe Campbell, *The Movable School Goes to the Negro Farmer* (Tuskegee: Tuskegee Institute Press, 1936), 160; Hersey, "'My Work is That of Conservation,'" 43–45.

35. Charles S. Johnson, *Shadow of the Plantation* (Chicago: University of Chicago Press, 1934), 13; Lawrence Elliott, *George Washington Carver: The Man Who Overcame* (Englewood Cliffs, NJ: Prentice-Hall, 1966), 107; "Movable Farmers Schools Last Week," n.d., clipping of unknown newspaper inside report of county agent, Bullock County, 1915, in ACES Records.

36. Rural farmers needed to do more to improve the totality of their lives, not simply work cotton to the exclusion of all else, he believed. Campbell's frustration in preaching diversification to farmers who ignored his message must have been similar to the dissatisfaction experienced by Carver. Years earlier, the scientist had told a local black landowner to consider a crop other than cotton for the following season. The farmer reportedly replied, "Son, I know all there is to know about farming. I've worn out three farms in my lifetime." Campbell, *The Movable School Goes to the Negro Farmer*, 109, 94; Hersey, "'My Work Is That of Conservation,'" 43–45.

37. The act built on the 1903 demonstration farm bill created in the aftermath of Knapp's supposed success against the boll weevil in Terrell, Texas. The Smith-Lever Act gave Knapp's loosely organized and often underfunded extension service federal institutional status. One of the bill's goals was to place an extension agent in the courthouse of every county in the country. Appropriations were based on the rural populations of states, the more farmers, the more funding. The federal government financially supported the system, but left the control of the workforce to the state land grant schools. The resulting Cooperative Extension Service notably failed to make specific contributions for black southern industrial schools. During the debates over the Smith-Lever bill, an amendment was added guaranteeing equal funding for black A&M institutions, but Georgia Senator Hoke Smith, the Senate sponsor of the legislation, bitterly opposed the amendment and it failed. The black extension service would continue to be separate and unequal, supported with less federal funding and still reliant on philanthropic donations. For the "white" state farm schools at least, the Smith-Lever Act fundamentally changed their mission, their connection to formal political bodies, and, most importantly, their institutional financial health. Pete Daniel, "The Crossroads of Change: Cotton, Tobacco, and Rice Cultures in the Twentieth-Century South," *Journal of Southern History* 50, no. 3 (August 1984): 434; *Smith Lever Act*, U.S. Code Title 7, Chapter 341, et seq. (1914).

38. Richmond Y. Bailey, transcript of interview by Norwood Kerr, July 19, 1982, Auburn University Agricultural Alumni Association Oral Histories. Auburn University Special Collections.

39. M. B. Ivy, "A Thought for Bullock County," n.d., unknown newspaper, clipping found inside report of county agent, Bullock County, 1915, in ACES Records.

40. Ibid.; Ivy, "Mithell Ivey [*sic* both names] Sees Bright Future," unknown newspaper, clipping found inside report of county agent Bullock County, 1916, in ACES Records.

41. Report of County Agent, Barbour County, 1916, in ACES Records; Report of County Agent, Geneva County, 1916, in ACES Records; Report of County Agent, Russell County, 1916, in ACES Records.

42. Kathryn Holland Braund, "'Hog Wild' and 'Nuts': Billy Boll Weevil Comes to the Alabama Wiregrass," *Agricultural History* 63, no. 3 (Summer 1989): 20.

43. Report of County Agent, Covington County, 1916, ACES Records.

44. Report of County Agent, Bullock County, 1915, ACES Records.

45. The USDA and state extension services across the South had long offered peanuts and sweet potatoes to farmers as replacements to cotton. Carver's research built on this previous work. More about Carver's role in this research is discussed in chapter six. George W. Carver, "Twelve Ways to Meet the New Economic Conditions Here in the South," Tuskegee Experiment Station Bulletin, no. 33 (1917).

46. Lawrence Elliott's discussion of this meal is the only reference to it the author has found. Elliott cites no specific date for the event, though he places it in the context of a 1915 Carver bulletin. Elliott, *George Washington Carver*, 151, 153-54.

47. The Census considered towns with more than 2,500 residents to be "urban." Fred Shelton Watson, *Piney Wood Echoes: A History of Dale and Coffee Counties, Alabama* (Elba, AL: Elba Clipper, 1949), 105-7; Braund, "'Hog Wild' and 'Nuts,'" 18.

48. There is no way to check the accuracy of the 60 percent crop loss figure. Though there were places in the South that experienced losses this heavy during one season, it was rare. This figure could have been a symptom of the weevil myth itself. Braund, "'Hog Wild' and 'Nuts,'" 20; Fred S. Watson, *Coffee Grounds: A History of Coffee County, Alabama, 1841-1970* (Anniston, AL: Higginbotham, 1970), 197.

49. Braund, "'Hog Wild' and 'Nuts,'" 22; USDA Bureau of Agricultural Economics, "Statistics on Cotton and Related Data," Statistical Bulletin 99 (June 1951), 34.

50. Watson, *Coffee Grounds*, 105; Braund, "'Hog Wild' and 'Nuts,'" 27.

51. Watson, *Coffee Grounds*, 105; Braund, "'Hog Wild' and 'Nuts,'" 27; Report of County Agent, Coffee County, 1916, in ACES Records, Auburn University, Auburn, Alabama. It may have been Pittman's idea that Sessions direct Baston to plant peanuts. The agent worked closely with Enterprise business interests. He wrote in 1916 that merchants "prefer to do business with them [agents], and often invite us to call on their regular customers." The move to peanuts was not relegated to the southeastern Alabama counties under study here, but the region was without a doubt the heart of the peanut boom. Two peanut researchers with the USDA claimed that total peanut acreage in the South doubled to two million acres from 1916 to 1917. H. S. Bailey and J. A. LeClerc, "The Peanut, A Great American Food," *Yearbook of the United States Department of Agriculture, 1917* (Washington: Government Printing Office, 1918), 289.

52. Report of County Agent, Coffee County, 1916, in ACES Records, ; Braund, "'Hog Wild' and 'Nuts,'" 27; *Thirteenth Census of the United States Taken in the Year 1910*, vol. 6: Agriculture, 1909 and 1910 (Washington, D.C.: Government Printing Office, 1913); *Fourteenth Census of the United States Taken in the Year 1920*, vol. 6, part 2: Agriculture (Washington, D.C.: Government Printing Office, 1922).

53. George W. Carver, "How to Grow the Peanut and 105 Ways of Preparing It for Human Consumption," Tuskegee Experiment Station Bulletin, no. 31, June 1915; H. S. Bailey and J. A. LeClerc, "The Peanut, A Great American Food," *Yearbook of the United States Department of Agriculture, 1917* (Washington: Government Printing Office, 1918), 289, 293.

54. Braund, "'Hog Wild' and 'Nuts,'" 23-24; Watson, *Coffee Grounds*, 105; Watson, *Piney Wood Echoes*, 108.

55. Report of County Agent, Bullock County, 1916, in ACES Records; Report of County Agent, Covington County, 1915, in ACES Records.

56. Report of County Agent, Dale County, 1919, in ACES Records.

57. Alyce Billings Walker, ed., *Alabama: A Guide to the Deep South*, new rev. ed. (New York: Hastings House, 1975), 330; Watson, *Coffee Grounds*, 105; Braund, "'Hog Wild' and 'Nuts,'" 27; Report of County Agent, Geneva County, 1917, in ACES Records; Report of County Agent, Barbour County, 1918, in ACES Records; Report of County Agent, Barbour County, 1918, in ACES Records; W. M. Welch to Farmers of Macon County, November 25, 1918, in Report of County Agent, Macon County, 1918, in ACES Records; Report of County Agent, Bullock County, 1918, in ACES Records.

58. Montgomery *Advertiser*, October 22, 1919, clipping in Report of County Agent, Geneva County, 1919, in ACES Records.

59. Ibid.; Watson, *Piney Wood Echoes*, 166; Watson, *Coffee Grounds*, 105, 107.

60. Walter M. Grubbs to George Washington Carver, November 30, 1919, in George Washington Carver papers, microfilm edition, reel 6; Walter M. Grubbs to George Washington Carver, December 12, 1919, in George Washington Carver papers, microfilm edition, reel 6; Walker, *Alabama*, 330; Watson, *Piney Wood Echoes*, 166; Hersey, "'My Work Is that of Conservation,'" 373–74.

61. Montgomery *Advertiser*, December 13, 1919, , p. 4.

CHAPTER 6

1. *Thirteenth Census of the United States Taken in the Year 1910*, vol. 6: Agriculture, 1909 and 1910 (Washington, D.C.: Government Printing Office, 1913); *Fourteenth Census of the United States Taken in the Year 1920*, vol. 6, part 2: Agriculture (Washington, D.C.: Government Printing Office, 1922).

2. *Thirteenth Census of the United States Taken in the Year 1910*, vol. 6: Agriculture, 1909 and 1910 (Washington, D.C.: Government Printing Office, 1913); *Fourteenth Census of the United States Taken in the Year 1920*, vol. 6, part 2: Agriculture (Washington, D.C.: Government Printing Office, 1922); Agricultural Census, 1925 (Washington, D.C., Government Printing Office, 1927); *Fifteenth Census of the United States Taken in the Year 1930*, agriculture vol. (Washington, D.C.: Government Printing Office, 1932).

3. *Fifteenth Census of the United States Taken in the Year 1930*, Aagriculture Volume vol. (Washington, D.C.: Government Printing Office, 1932).

4. Ibid.

5. Ibid. *Twelfth Census of the United States: 1900* (Washington, D.C.: Government Printing Office, 1902); *Fifteenth Census of the United States Taken in the Year 1930*, agriculture vol. (Washington, D.C.: Government Printing Office, 1932); Braund, "'Hog Wild' and 'Nuts,'" 33.

6. *Thirteenth Census of the United States Taken in the Year 1910*, vol. 6: Agriculture, 1909 and 1910 (Washington, D.C.: Government Printing Office, 1913); *Fourteenth Census of the United States Taken in the Year 1920*, vol. 6, part 2: Agriculture (Washington, D.C.: Government Printing Office, 1922); Agricultural Census, 1925 (Washington, D.C., Government Printing Office, 1927); *Fifteenth Census of the United States Taken in the Year 1930*, agriculture vol. (Washington, D.C.: Government Printing Office, 1932).

7. J. P. Wilson, "Cotton Acreage," February 1, 1919, typed article in Report of County Agent, Covington County, 1919, ACES Records.

8. Ibid.

9. USDA Bureau of Agricultural Economics, "Statistics on Cotton and Related Data," Statistical Bulletin 99 (June 1951), 34.

Growers were dazzled by even the prospect of a bumper cotton crop; high prices blinded them to potential losses and the cyclical debt that was part and parcel of cotton farming for all but the wealthiest planters. J. P. Wilson, the agent who promised growers that cotton would make them "poor and ignorant," admonished landowners and tenants to "Forget the idea of getting rich in one year." Few listened. In their year-end reports, local agents bemoaned farmers' attraction to high prices. At the end of the 1919 season, Houston County agent J. H. Witherington lamented that "Cotton acreage was increased this year" despite the boll weevil reducing farmers' yields. "We will have a large acreage next year owing to the present high prices," he predicted. Though he tried to stop farmers from growing more cotton, once they had made up their minds, Witherington found there was nothing to do except try to help them get the greatest cotton yield they could. Instead of advising demonstrators on peanuts and potatoes, he responded to questions about weevil poisons. J. P. Wilson, "Cotton Acreage," February 1, 1919, typed article in Report of County Agent, Covington County, 1919, ACES Records. Report of County Agent, Houston County, 1919, ACES Records. Report of County Agent, Houston County, 1920, ACES Records.

10. Report of County Agent, Houston County, 1919, ACES Records. Report of County Agent, Houston County, 1920, ACES Records. *Thirteenth Census of the United States Taken in the Year 1910*, Volume VI: Agriculture, 1909 and 1910 (Washington, D.C.: Government Printing Office, 1913); *Fourteenth Census of the United States Taken in the Year 1920*, Volume VI, part 2: Agriculture (Washington, D.C.: Government Printing Office, 1922); Agricultural Census, 1925 (Washington, D.C., Government Printing Office, 1927); *Fifteenth Census of the United States Taken in the Year 1930*, Agriculture Volume (Washington, D.C.: Government Printing Office, 1932). The author figured value per acre by first computing bales and bushels per acre from census data, then multiplying the yield figure with the price. The author figured the dollar value per acre figures assuming cotton bales were five-hundred pounds and peanut bushels as twenty-two-pounds.

11. Report of County Agent, Houston County, 1919, ACES Records; Report of County Agent, Houston County, 1920, ACES Records.

12. On systemic problems of the USDA and extension services, see Helms, "Just Lookin' for a Home," chapters 3–4; Sanders, *Roots of Reform*, chapter 9.

13. Martin, *Demonstration Work*, 8, 9.

14. "Caterpillar Cause of Much Complaint," Montgomery *Advertiser*, August 14, 1911, newspaper clipping in Hinds Papers, box 1, folder 10; "Duggar Investigates Alabama Cotton Pests," newspaper clipping [no paper name], September 1, 1911, in Hinds Papers, box 1, folder 6. Knapp quoted in Bailey, *Seaman A. Knapp*, 213.

15. Warren E. Hinds, "Cotton Dusting Record: 1919," in Hinds Papers, box 1, folder 4; Martin, *Demonstration Work*, 32.

16. "Alabama County Statistical Sheet," Hinds Papers, box 1, folder 16.

17. See county agents reports beginning 1918, ACES Records. See also Harold Woodman, *New South, New Law: The Legal Foundations of Credit and Labor Relations in the Postbellum Agricultural South* (Baton Rouge: Louisiana State University Press, 1995). Black agents were instructed that they "need not report" on these new forms.

18. Report of County Agent, Barbour County, 1918, ACES Records; Report of County Agent, Covington County, 1919, ACES Records; Report of County Agent, Dale County, 1919, ACES Records; Report of County Agent, Bullock County, 1919, ACES Records.

19. Report of County Agent, Geneva County, 1918, ACES Records; Report of County Agent, Geneva County, 1919, ACES Records; Report of County Agent, Macon County, 1918, ACES Records.

20. See Mark D. Hersey, *My Work Is That of Conservation: An Environmental Biography of George Washington Carver* (Athens: University of Georgia Press, 2011).

21. Report of County Agent, Barbour County, 1918, ACES Records.

22. Karen J. Ferguson, "Caught in 'No Man's Land': The Negro Cooperative Demonstration Service and the Ideology of Booker T. Washington, 1900–1918," *Agricultural History* 72, no. 1 (Winter 1998): 29, 44. Historian Robert J. Norrell has concluded that Tuskegee's "efforts to improve farming practices apparently benefited black farmers in Macon County only marginally." Norrell, *Reaping the Whirlwind*, 23.

23. The Black Belt counties of southeastern Alabama had the direst numbers. Eight of the thirteen counties under study ranked in the bottom half of the state's agricultural income ranking. USDA, Bureau of Agricultural Economics, "Statistics on Cotton and Related Data," Statistical Bulletin 99 (June 1951), 61.

CHAPTER 7

1. Maurice L Friedman, "The World War Years—Coming of the Boll Weevil, 1910–1945," in *Cotton to Kaolin: A History of Washington County, Georgia, 1784–1989*, ed. Mary Alice Jordan (Sandersonville, GA: Washington County Historical Society, 1989), 80, 83.

2. Arthur Franklin Raper, *Preface to Peasantry: A Tale of Two Black Belt Counties* (Chapel Hill: University of North Carolina Press, 1936).

3. Ibid. Willard Range, *A Century of Georgia Agriculture, 1850–1950* (Athens: University of Georgia Press, 1954), 175.

4. In an attempt to discover which farmers were managing a profit despite these high-bulk line prices, the investigators examined farmers' costs in terms of their yield. They found that the higher the yield farmers squeezed from their land, the lower their cost of production. In other words, if a farmer managed only 172 pounds of lint cotton per acre, it cost 11.6 cents per pound to produce, but if they increased their yields to 456 pounds per acre, the cost dropped to 7.5 cents per pound. Paradoxically, in order to meet these higher production costs, farmers had to spend money, labor, and time to improve their yields, so that their costs might decline. For most, this meant heavy fertilizer use. Over the one-hundred-year history of cotton production in Georgia, landowners had rarely rotated fields to rest the soil, and this constant cotton production had robbed the land of its nutrients. Georgians consistently increased their use of fertilizer each year in a desperate attempt to replace the nutrients that generations of cotton

crops had taken from the earth. This fertilizer came at a high cost, however, which cut into profit margins. Cotton farmers were caught in a pinch: if they wanted to lower their production costs, they needed more capital in order to make their land more productive. Though for a decade these growers had been losing money, despite increasing their cotton production, things were about to get worse. USDA Bureau of Agricultural Economics, "Statistics on Cotton and Related Data," Statistical Bulletin 99 (June 1951), 136; "Stop the Losses in Farming," Georgia State College of Agriculture, vol. 8, no. 33, bulletin 211 (1920), 10, 14, 15; Fite, *Cotton Fields No More*, 28.

5. Dickson, *The Story of King Cotton*, 96. On the Georgia coast, a number of farmers grew long-staple Sea Island cotton, though it was a miniscule portion of the state's overall cotton production. The boll weevil's arrival there destroyed all hopes of growing that breed profitably. Sea Island cotton demanded a long growing season, which meant that as its bolls ripened in late summer, when boll weevils were at full strength. Sea Island cotton disappeared almost overnight.

6. Range, *A Century of Georgia Agriculture*, 175.

7. Stephen J. Karina, *The University of Georgia College of Agriculture: An Administrative History, 1785–1985* (Athens: University of Georgia, 1989), 123–24. The Board of Trustees chose Soule over several other candidates, including University of Georgia alumnus Ulrich B. Phillips, who refused the job offer.

8. Andrew M. Soule, "Preparing Georgia for the Advent of the Boll Weevil," n.d., Andrew M. Soule Papers, University Archives, University of Georgia, hereafter Soule Papers; Soule, "The Mexican Cotton Boll Weevil," n.d., in folder "Lectures, Cotton School, January 6–17, 1908," Soule Papers; "The Cotton School" (Athens: Georgia State College of Agriculture, November 1907), vol. 8, no. 3, serial no. 77, Soule Papers.

9. J. Phil Campbell, "Extension Service Report for Georgia, 1915–1916" (Athens: Georgia State College of Agriculture, August 1916), vol. 4, no. 1, bulletin 110.

10. Soule, "Preparing Georgia for the Advent of the Boll Weevil," n.d., Soule Papers; Central of Georgia Railway, "The Boll Weevil is Coming! What Are You Going To Do About It?" (Savannah: Central of Georgia Railway, 1914).

11. Invitation for "Annual Meeting of the Georgia Dairy and Live Stock Association, Georgia Breeders' Association, State Horticultural Society and A Boll Weevil Conference," 1917, Soule Papers.

12. J. Phil Campbell, "Farmers' Co-operative Demonstration Work in Georgia" (Athens: Georgia State College of Agriculture, February 1914), vol. 2, no. 8; Soule, "Preparing Georgia for the Advent of the Boll Weevil," n.d., Soule Papers.

13. "Announcements of Short Courses" (Athens: Georgia State College of Agriculture, November 1916), vol. 4, no. 3, bulletin 112; J. G. Oliver, "Starve the Boll Weevil" (Athens: Georgia State College of Agriculture, September 1919), vol. 8, no. 10, bulletin 188; W. E. Hinds, "Beating the Boll Weevil," in "Proceedings of Georgia State Horticultural Society, Georgia Market Conference, Georgia Breeder's Association and Georgia Dairy and Live Stock Association" (Athens: Georgia State College of Agriculture, June 1916), vol. 4, no. 16, p. 71–81; Frank C. Ward, "Poison Boll Weevils" (Athens: Georgia State College of Agriculture, January 1920), vol. 8, no. 18, bulletin 196.

14. Dickerson et al., eds., *Boll Weevil Eradication in the United States through 1999*, 598, 601, 604; Soule, "Annual Report of the President, State College of Agriculture and the Mechanic Arts, 1921–1922" (Athens: Georgia State College of Agriculture, June 1922), vol. 10, no. 15, bulletin 254.

15. Harvie Jordan to John Judson Brown, May 4, 1922, in John Judson Brown Papers, Southern Historical Collection, University of North Carolina at Chapel Hill, hereafter Brown Papers, box 1, folder 22.

16. Ibid.

17. J. J. Brown to J. L. Smith, February 11, 1922, in Brown Papers, box 1, folder 18.

18. Ibid.

19. Range, *A Century of Georgia Agriculture*, 141. For a longer discussion on the cooperative movement in the South, see Woodward, *Origins of the New South*; Ayers, *Promise of the New South*.

20. J. J. Brown to J. L. Smith, February 11, 1922, in Brown Papers, box 1, folder 18.

21. Report of the Commissioner of Agriculture, 1922, in Brown Papers, box 1, folder 23.

22. J. J. Brown to J. Polk Brown, July 13, 1922, in Brown Papers, box 1, folder 25.

23. Ibid.

24. "John William Firor, Sr. 1887–1956," John William Firor Papers, Special Collections, Perkins Library, Duke University, hereafter Firor Papers; Andrew M. Soule to John Firor, April 10, 1919, Firor Papers. Firor's daughter, historian Anne Firor Scott, has written about her youth and family in Anne Firor Scott. "My Twentieth Century: Leaves From a Journal," *Southern Cultures* 9, no.1 (Spring 2003): 62–78.

25. John Firor to Robert Schmidt, September 25, 1919, Firor Papers.

26. Firor's job led him to be concerned about the boll weevil's movements across the South. He wrote to one farmer asking if "the boll weevil has done great damage to cotton this year." In another he asked his brother, "To what extent has the boll weevil damaged the cotton in your neck of the woods and what effect has it had on the morale of large planters?" John Firor to Paul Moore, October 2, 1919, Firor Papers; John Firor to Guy Firor, September 3, 1919, Firor Papers; Robert Schmidt to John Firor, October 4, 1919, Firor Papers.

27. John Firor to M. C. Gay, November 20, 1919, Firor Papers.

28. United States Railroad Administration to "Sir," December 12, 1919, Firor Papers; John Firor to William H. Barrett, December 11, 1919, Firor Papers.

29. John Firor to Mary Valentine Ross, October 1919, Firor Papers.

30. "Georgia's Record of Farming, A Pledge of Prosperity," *Atlanta Journal*, March 12, 1930; Soule, "Whipping Billy Boll Weevil," n.d., Soule Papers.

31. Thomas P. Janes, *Annual Report of Thom P. Janes, Commissioner of Agriculture for the State of Georgia, for the Year 1875* (Atlanta: 1875), as quoted in Roger Ransom and Richard Sutch, "Tenancy, Farm Size, Self-Sufficiency, and Racism: Four Problems in the Economic History of Southern Agriculture," Southern Economic History Project, Working Paper Series no. 8 (Institute of Business and Economic Research, University of California, Berkeley, April 1970), 75–76.

32. "In Lieu of Something Better," *American Life Histories: Manuscripts from the Federal Writers' Project, 1936–1940*, published electronically by the Library of Congress, available at http://lcweb2.loc.gov/ammem/wpaintro/wpahome.html; hereafter *American Life Histories*.

33. Ibid.

34. Ibid.

35. "Tricked By Gypsies," *American Life Histories*.

36. Ibid.

37. Ibid.

38. Ibid.

39. Range, *A Century of Georgia Agriculture*, 259. For a longer discussion of the obstacles to black landownership, see Jay R. Mandle, *The Roots of Black Poverty: The Southern Plantation Economy After the Civil War* (Durham: Duke University Press, 1978); Gavin Wright, *Old South, New South: Revolutions in the Southern Economy since the Civil War* (New York: Basic Books, 1986); Ransom and Sutch, *One Kind of Freedom*.

40. Hope T. Eldridge and Dorothy Swaine Thomas, "Demographic Analyses and Interrelations," in *Population Redistribution and Economic Growth: United States, 1870–1950*, ed. Simon Smith Kuznets, vol. 3 (Philadelphia: American Philosophical Society, 1964), 16–17.

41. Raper, *Preface to Peasantry*, 204, 201–2.

42. Ibid., 201–2, 205, 206, 209.

43. *Fifteenth Census of the United States Taken in the Year 1930*; *Sixteenth Census of the United States Taken in the Year 1940*.

44. Numan V. Bartley, *The Creation of Modern Georgia*, 2nd ed. (Athens: University of Georgia Press, 1990), 130.

45. Steven Hahn, *A Nation Under Our Feet: Black Political Struggles In The Rural South, From Slavery to The Great Migration* (Cambridge: Belknap Press of Harvard University Press, 2003), 466. Baraka is quoted in Farah Jasmine Griffin, *"Who Set You Flowin'?": The African-American Migration Narrative* (New York: Oxford University Press, 1995), 18.

46. Posey Oliver Davis, "Migration in Relation to Alabama Agriculture, 1940," typed manuscript in Alabama Collection, Auburn University Special Collections.

47. Ibid.

48. John Egerton, *Speak Now Against the Day: The Generation Before the Civil Rights Movement in the South* (New York: Knopf, 1994), 50; Nancy McLean, *Behind the Mask of Chivalry: The Making of the Second Ku Klux Klan* (New York: Oxford University Press, 1994), 183.

49. Whatley and Wright, "Black Labor in the American Economy since Emancipation," 76–79.

50. Griffin, *"Who Set You Flowin'?,"* 22.

51. Leadbelly, "Boll Weevil" (audio recording), Library of Congress Recordings (LC 273-A-1 and LC 135-A).

52. Paul Oliver, *Blues Fell This Morning: Meaning in the Blues*, 2nd ed. (Cambridge: Cambridge University Press, 1990), 16, 18; Harris, *Deep Souths*, 251, 284.

53. Blind Willie McTell, "Boll Weevil" (audio recording), Library of Congress Recordings (LC AFS L51).

54. Buster "Bus" Ezell, "Boll Weevil," from "'Now What a Time': Blues, Gospel, and the Fort Valley Music Festivals" (audio recording), American Folklife Center, Library of Congress Recordings.

55. Woody Guthrie, "Boll Weevil Blues," in *Woody Guthrie Sings Folk Songs* (audio recording), Smithsonian Folkways (40007); Cisco Houston, "Boll Weevil," in *American Roots: A His-*

tory of American Fold Music (audio recording), Disky (248612); Carl Sandburg, "Boll Weevil Song," in *New Songs from the American Songbag* (audio recording), Lyricord (LL 4).

56. For background on how the chicken industry supplanted cotton in northern Georgia, see Monica Gisolfi, "From Cotton Farmers to Poultry Growers: The Rise of Industrial Agriculture in North Georgia, 1914–1975" (Ph.D. dissertation, Columbia University, 2007), and Gisolfi, "From Crop Lien to Contract Farming: The Roots of Agribusiness in the American South, 1929–1939," *Agricultural History* 80:2 (Spring 2006).

CONCLUSION

1. Dan Charles, "Boll Weevil Eradication," report on *Morning Edition*, National Public Radio, July 7, 2003. An archived audio recording of this program was accessed by the author on the Internet, November 8, 2004, at http://www.npr.org/templates/story/story.php?storyId=1321881; emphasis added.

2. Ibid.

3. Ibid.

4. Vershal Hogan, "Boll Weevils Almost Gone," *Natchez Democrat* (Mississippi), May 10, 2010 http://www.natchezdemocrat.com/news/2010/may/10/boll-weevils-almost-gone/ (accessed May 23, 2010); Betsy Blaney, "Bugged Farmers Beat Boll Weevil," *Memphis Commercial Appeal*, November 1, 2009, http://www.allbusiness.com/agriculture-forestry-fishing-hunting/agriculture-crop/13367188-1.html (accessed May 23, 2010).

5. Southern Regional Committee of the Social Science Research Council, "Problems of the Cotton Economy: Proceedings of the Southern Social Science Research Conference" (Dallas: Arnold Foundation, 1936), in Howard W. Odum Papers, Southern Historical Collection, University of North Carolina at Chapel Hill, series 2.1, folder 628.

6. Gilbert C. Fite, "Southern Agriculture Since the Civil War: An Overview," in George L. Robson Jr. and Roy V. Scott, eds. *Southern Agriculture Since the Civil War: A Symposium* (Santa Barbara: McNally & Lotin, West, 1979), 16.

7. William Neuman and Andrew Pollack, "Farmers Cope With Roundup-Resistant Weeds," *New York Times*, May 3, 2010, http://www.nytimes.com/2010/05/04/business/energy-environment/04weed.html (accessed May 23, 2010).

ACKNOWLEDGMENTS

I've been thinking about the boll weevil for a long time, and I'm the kind of person who can't think to or by himself. So I came to depend on the knowledge, advice, criticism, humor, and love of family, friends, and complete strangers.

I first became interested in the rural South as a junior at DePauw University, in a class taught by John Dittmer. John has been a helpful critic, unflagging supporter, and steadfast friend since then. His teaching and scholarship are models for my own. He is why I became a historian. I owe him this first official thank-you.

This book took its first breaths in seminars at the University of North Carolina at Greensboro, where I had not only institutional support, but also a bunch of bright people around me shaping my thinking. I owe a great deal to Bill Link, Bill Blair, and Chuck Holden for teaching me how to be a historian. Since those years, Chuck has become my go-to critic. I don't consider anything I write truly finished until Chuck has given it clearance. We've also traveled together throughout the South, which has so far resulted in only one ugly run-in with the Mississippi State Police. He is a great friend and has made this book better.

At the University of Georgia passionate folks interested in southern history surrounded me and there are many days I wish I could transport myself back there. I've kept Paul Sutter's heavily inked copies of my papers and essays on my desk for longer than I'd like to admit, so he's taught me as much since I left Athens as he did while I was there. Bryant Simon lived around the corner from me for my last two years of school and, as a result, heard more about my work than he wanted to. But he never grew tired of my nagging and prodding, and this book is in part the result of questions and comments he offered in our yards and on the sidewalks between, usually with his kids perched atop or hanging from at least one of us. Bryant, Ann Marie, Benjamin, and Eli made my life in Athens a great joy. They fed my family, kept us laughing, found us housing, employed us, and just brought us in. I can never repay that debt.

James Cobb deserves his own paragraph. He offered encouragement for this project at precisely the moments I needed it, and criticism when and where I deserved it most. There are few people who are at once as serious and as funny as Jim and there is no one I more enjoy talking with about the South.

There are others who have commented upon, encouraged, and set ablaze my interpretations of the boll weevil. These people did so in classrooms, conference sessions, e-mail exchanges, and far less formal settings. They each deserve a formal thank-you, but they'll have to settle for their name in a list: Stephen Brain, Mark Cioc, Kathleen Clark, Peter Coclanis, Pete Daniel, Monica Gisolfi, Sarah Gregg, Valerie Grim, Steven Hahn, Shane Hamilton, John Hayes, Elizabeth Herbin, Mark Hersey, Mark Huddle, Tammy Ingram, Vasco Lorbner, Chris Manganiello, Alan Marcus, Peter Messer, Ichiro Miyata, Adrienne Petty, Jason Phillips, Debra Reid, Susan Rensing, Drew Swanson, Bert Way, Jeannie Whayne, and the anonymous reviewers for the University of Chicago Press. My time writing this book would have been miserable without Alex Macaulay, Brant Rumble, and Kevin Waltman. It would have been uninspired without Patterson Hood and Dave Marr.

Archivists and librarians are the gatekeepers of history and I heartily thank all who helped me dig through their books and manuscripts. Bill Marshall at the University of Kentucky, Mattie Sink of Mississippi State University, and Dwayne Cox at Auburn went above and beyond their duties to help me. I'm also indebted to the staff of the University of Chicago Press, especially Robert Devens, whose support of this book was unwavering even as deadlines sailed by, Anne Summers Goldberg, who has kept me on track as the finish line came into sight, and Michael Koplow, whose attention to detail is inspiring. I also thank *Environmental History* and *Agricultural History* for granting permission to publish sections of this work that first appeared in their pages.

Finally, I thank my family, who put up with the most. My parents, Phil Giesen and Barb Giesen, have supported me in ways I can never repay and in ways that only now, as a parent myself, do I begin to fully understand. I hope they know that I appreciate what they've done for me. Bill and Jan Marshall have gone to unnecessary lengths to encourage my pursuits in the academy, at home, and everywhere in between. My sisters, Betty Scalia and Katie Giesen, are remarkable women who continue to inspire me, though from way too far away. I constantly miss them. Having Jeff Scalia as my brother-in-law has paid off precisely once: when he drew a map and all the graphs for this book. My kids, Walter and Eleanor Giesen, now that I think about it, didn't do too much to help me finish this book. They owe me.

Anne Marshall is a joy to live and work with. She continually astonishes me with her accomplishments as a scholar, partner, mother, and friend. She's made this book—and me—better, and for that I thank her.

INDEX